Explainable AI

With the increasing use of Deep Learning systems across various industries, there is a growing need to make their decision-making processes more understandable and transparent. Regulatory requirements now demand clarity, and users and stakeholders want to know how AI systems work. This textbook addresses these needs by providing a detailed guide on integrating Explainable AI (XAI) into the Deep Learning Operations (DLOps) pipeline. By doing so, organizations can implement Continuous Integration (CI) and Continuous Deployment (CD) practices effectively.

Explainable AI: Building Trustworthy Deep Learning Systems focuses on how to incorporate XAI methods, tools, and techniques to clarify Machine Learning decisions. It explores applications in fields such as healthcare, defense, human activity recognition, and object identification. The book offers practical advice on embedding XAI tools throughout the lifecycle of Deep Learning systems, covering topics such as Explainability and Interpretability, Deep Learning Operations (DLOps), and Machine Learning Operations (MLOps). It also includes real-world examples, challenges, and solutions.

This textbook is ideal for undergraduate and graduate students studying computer science, electronics and communication engineering, electrical and electronics engineering, and AI/ML specializations. It is particularly suited for courses such as Healthcare Analytics, Responsible and Explainable AI, AI Internals in Cyber-Physical Systems, AI Security Analytics, and Human–Computer Interaction with XAI.

Researchers working in the area of Trustworthy AI, Responsible AI, and Explainable AI can use this book as a primary source as it contains code implementations with metrics calculations in detail. Moreover, professionals in IT indsutries applying AI in software development such as DevOps and Deep Learning architecture development with XAI integrations in systems engineering and industrial engineering will find it a highly valuable development guide.

For those adopting the textbook for courses, a solutions manual and PowerPoint slides are available.

Prospects in Smart Technologies

Mohammad M. Banat, Series Editor
Mohammad M. Banat, Jordan University of Science and Technology, Irbid, Jordan
banat@just.edu.jo
Sara Paiva, Instituto Politécnico de Viana do Castelo, Viana do Castelo, Portugal
sara.paiva@estg.ipvc.pt

Emerging Technologies for Sustainable and Smart Energy
Edited by Anirbid Sircar, Gautami Tripathi, Namrata Bist, Kashish Ara Shakil, and Mithileysh Sathiyanarayanan

CYBORG
Human and Machine Communication Paradigm
Kuldeep Singh Kaswan, Jagjit Singh Dhatterwal, Anupam Baliyan, and Shalli Rani

Emerging Technologies and the Application of WSN and IoT
Smart Surveillance, Public Security, and Safety Challenges
Edited by Shalli Rani and Himanshi Babbar

WSN and IoT
An Integrated Approach for Smart Applications
Edited by Shalli Rani and Ashu Taneja

Immersive Technologies
Navigating the Impacts, Challenges, and Opportunities
Sagaya Aurelia

Health 5.0
Concepts, Challenges, and Solutions
Hiren Kumar Thakkar, Chintan Bhatt, Victor C. M. Leung, and Ilangko Balasingham

Integrating AI Techniques into the Design and Development of Smart Cyber-Physical Systems
Defense, Biomedical, Infrastructure, and Transportation
D. Jeya Mala, Anto Cordelia Tanislaus Antony Dhanapal, Saurav Sthapit, and Anita Khadka

Explainable AI
Building Trustworthy Deep Learning Systems
D. Jeya Mala and Subramaniam Ganesan

For more information on this series, please visit: https://www.routledge.com/Prospects-in-Smart-Technologies/book-series/CRCPST

Explainable AI

Building Trustworthy Deep Learning Systems

D. Jeya Mala

and

Subramaniam Ganesan

CRC Press
Taylor & Francis Group
Boca Raton London New York

CRC Press is an imprint of the
Taylor & Francis Group, an **informa** business

Designed cover image: Shutterstock

First edition published 2027
by CRC Press
2385 NW Executive Center Drive, Suite 320, Boca Raton FL 33431

and by CRC Press
4 Park Square, Milton Park, Abingdon, Oxon, OX14 4RN

CRC Press is an imprint of Taylor & Francis Group, LLC

ISBN: 978-1-041-04687-5 (hbk)
ISBN: 978-1-041-04851-0 (pbk)
ISBN: 978-1-003-63024-1 (ebk)

DOI: 10.1201/9781003630241

Typeset in Times
by SPi Technologies India Pvt Ltd (Straive)

Access the Instructor and Student Resources: www.Routledge.com/9781041046875

Dedication

Dedicated to all whose aim is to develop trustworthy AI models for real-time application development

Contents

Foreword

I am happy to write the foreword to this book. It is clear from the table of contents that the authors are experts in Artificial Intelligence. They have explained the new concepts with examples. This is an evolving field. There are not many books on this topic. The table of contents flows nicely and covers topics in a clear sequence. The table of contents reveals the technical knowledge of the authors. The book starts with an introduction to XAI, integration of XAI, and DLOps. This book covers model validation and interesting applications. Every chapter has many references for clear understanding and research.

Graduate and undergraduate students, along with practicing engineers, will find the book very useful. This book is really appealing to researchers and application developers. I kept reading this book till the end since it had good clarity.

Prof. S. Manimurugan
M.Eng., Ph.D, PDF (USA), MISTE, SMIEEE,
Professor, Department of Information Technology
Faculty of Computers and Information Technology
University of Tabuk, Tabuk City, Saudi Arabia

Preface

Nowadays, real-world Machine Learning and Deep Learning applications are governed by the General Data Protection Regulation (GDPR) law, which states "the right of explanation for all individuals to obtain meaningful explanations of the logic involved" when automated decision-making takes place.

Advanced Deep Learning (DL) models are being developed for specific use cases such as healthcare, finance, education, insurance, manufacturing, etc. These AI models make "black-box" forecasts without explaining how they did so. The requirement for researchers, organizations, and regulators to comprehend how AI models make recommendations, forecasts, etc. led to "Explainable AI (XAI)."

Hence, the need for XAI in DL is not only for technical insights but also for trustworthy, ethical, operational, and strategic purposes.

Based on this insight, this book, *Explainable AI: Building Trustworthy Deep Learning Systems*, explores the integration of XAI principles into the challenges of managing DL workflows. DLOps, which is the specialized subset of MLOps focused on DL, deals with complex models, big data, and resource-intensive training and deployment processes.

It presents a detailed account of different types of XAI methods such as model-specific and model-agnostic methods with case studies. The book elaborates various XAI methods such as LIME, SHAP, GradCAM, GradCAM++, Counterfactual Explanations, and ProtopNet with samples. Further, this book provides experimental case studies on how these XAI methods are applied to diverse applications such as healthcare, defense, human activity recognition, object identification, etc.

By embedding such XAI methods into the DLOps lifecycle, organizations can achieve not only transparency and accountability but also ways to improve the DL models' reliability and operational effectiveness.

NEED FOR THIS BOOK

As industries increasingly adopt DL systems, there is growing demand for interpretable and explainable insights into their decision-making processes. Regulatory frameworks also require transparency, and end users and stakeholders seek to understand the working of AI systems. This book addresses these demands by providing a comprehensive overview of how to integrate XAI into the DLOps pipeline so that Continuous Integration (CI) and Continuous Deployment (CD) practices can be achieved.

READERS OF THIS BOOK

This book is designed especially for those at the intersection of AI, ML, DL, and operational systems, including:

- UG and PG students who have elective courses on Explainable AI and/or DevOPs/MLOps/DLOps in their curriculum.
- Deep learning practitioners who wish to make their DL models interpretable and accessible.
- DLOps engineers who maintain robust and explainable pipelines for Deep Learning workflows.
- Researchers exploring the synergy between XAI and DL operations (DLOps).

STRUCTURE OF THE BOOK

This book has the following divisions:

1. Foundations – Focuses on the introduction to XAI, Explainability vs. Interpretability, DLOps, differences between MLOps and DLOps, along with the theoretical underpinnings of explainability in deep neural networks.
2. Integration of XAI in DLOps – Explores practical guidance on how to embed XAI tools and techniques at each stage of the DLOps lifecycle, starting from model development through deployment and monitoring.
3. Applications and Case Studies – Deals with the real-world integration of XAI into DLOps workflows across Agentic AI- and LLM-based applications.

As a summary, this book reflects the contributions and insights of researchers, practitioners, and DL engineers who are working in the advanced fields of XAI, Deep Learning, and DLOps.

By providing the right blend of XAI and DLOps, this book helps readers navigate the challenges and opportunities of integrating explainability into DLOps. This will definitely pave the way not only for cutting-edge technology but also for transparent and accountable systems aligned with ethical and responsive AI systems.

For those adopting the textbook for courses, a solutions manual and PowerPoint slides are available (www.routledge.com/9781041046875).

Acknowledgments

At the outset, we wholeheartedly thank the Lord Almighty, who lifts up the poor from dust and who gave us the wisdom, knowledge, courage, and vision to complete this book successfully.

It is our pleasure to express with deep sense of gratitude to the management and leaders of Vellore Institute of Technology, Chennai, Tamil Nadu, India and Oakland University, USA for providing all the infrastructure support and necessary resources for the successful completion of this book.

We express ingeniously our whole-hearted thanks to thank our family and friends who encouraged us to take up and complete this task.

We would like to thank the following individuals for their contributions:

Dr. Rajkumar J. Bhojan, AI Researcher, Fidelity Investments, Smithfield, RI, USA

Mr. Aditya Guleria, Student, Vellore Institute of Technology, Chennai, India

Ms. Kashish, Student, Vellore Institute of Technology, Chennai, India

Mr. Jeevan Jyothi, Student, Vellore Institute of Technology, Chennai, India

Mr. Gouri Ramachandran, Student, Vellore Institute of Technology, Chennai, India

Mr. A. Pradeep Reynold, NSH Corporation Ltd., Saudi Arabia

About the Authors

Dr. D. Jeya Mala is currently a professor in the School of Computer Science and Engineering (SCOPE), Vellore Institute of Technology, Chennai, Tamil Nadu, India. She has more than 24 years of teaching and research experience and 4 years of industrial experience. As a member of the "National Work Group on Quantum Computing" formed by the Telecommunications Department, Government of India, she contributed to India's proposal on Quantum Computing for Future Networks at the Geneva Meet, Switzerland. She served as an expert evaluation committee member of AICTE-NEAT, Government of India, and MoE's Innovation Council, Government of India. To her research credit, she has one granted design patent, three published utility patents from IP, Government of India, one completed funded research project, 4 books, and more than 65 papers published in reputed refereed SCI- and Scopus-indexed journals, conferences, and book chapters. She is currently working on a funded quantum-based collaborative research project with Deakin University, Australia. She is a listee of *Who's Who* list of SEBASE repository of University College London, UK, for her research work in the area of search-based software engineering and a proud recipient of several laurels and awards, and member of IEEE, ACM, etc. Her research interests include artificial intelligence, quantum computing, DL and ML, XAI, healthcare analytics, software engineering, cybersecurity, and blockchain.

Dr. Subramaniam Ganesan, a professor in the Department of Electrical & Computer Engineering (ECE), Oakland University, Rochester, Michigan, USA. He is a senior member of IEEE, former IEEE Computer Society Distinguished Visiting Speaker, IEEE Region for 4 Technical Activities member, and fellow of ISPE. He has received the Lifetime Achievement Award from ISAM, the Lloyd L. Withrow Distinguished Speaker Award from SAE, the Best Paper Award from ISAM, the Best Teacher Award from ASEE, and similar accolades from Oakland University. He is the editor-in-chief of the *International Journal of Embedded Systems & Computer Engineering*, as well as the *International Journal of Sensors & Applications*. He has been the session organizer of the "Systems Engineering" panel at the SAE World Congress for the past 15 years. More details can be viewed on the home page at: www.secs.oakland.edu/~ganesan. His research interests are in real-time systems, parallel architectures, mobile computing, automotive embedded systems, and signal processing. He holds several patents in embedded systems.

Part 1

Foundations on XAI

1 Introduction to XAI Taxonomy, Ethics, and Policy

1.1 INTRODUCTION TO INTELLIGENT SYSTEMS

Advanced Artificial Intelligence (AI) models developed for specific use cases such as healthcare, finance, education, insurance, etc. provide "black-box" forecasts without explaining how they did it. As real-time systems analytics need to give crucial decision-making, the deployment of Machine Learning (ML) or Deep Learning (DL) models for such critical settings needs to be examined for their trustworthiness. The application of ML and DL algorithms to real world problems is a boon to the development of smart systems. Several applications under real-time domains such as remote patient monitoring systems, cancer diagnostics, cardiovascular disease prediction, precision nutrition in healthcare; smart transportation systems, smart route planning, smart agriculture, etc. in Smart cities; smart defense applications, smart communication systems, etc. are all using ML and DL based analytics to make appropriate decisions.

1.2 MACHINE LEARNING AND DEEP LEARNING MODELS – AN INTRODUCTION

According to Arthur Samuel,

> Machine Learning algorithms enable the computers to learn from data, and even improve themselves, without being explicitly programmed.

Machine Learning (ML) algorithm allows software applications to become more accurate in predicting outcomes without being explicitly programmed.

The basic premise of ML is to build algorithms that can receive input data and use statistical analysis to predict an output while updating outputs as new data becomes available. The generic process flow of the ML/DL model is given in Figure 1.1.

To analyze the given domain, preparation of a problem statement is essentially the first step to start with. Based on the problem definition, the next task is to collect the corresponding data to train the ML/DL model. If the dataset collected is very small, that may lead to overfitting at the time of prediction as the model trained on less amount of data. Now, it is essential to augment the existing dataset with appropriate data by means of various data augmentation techniques such as SMOTE etc.

Once the dataset is ready, it is time to preprocess the data as the data may have noise and bias in it, which will affect the decision made by the model. Using data

DOI: 10.1201/9781003630241-2

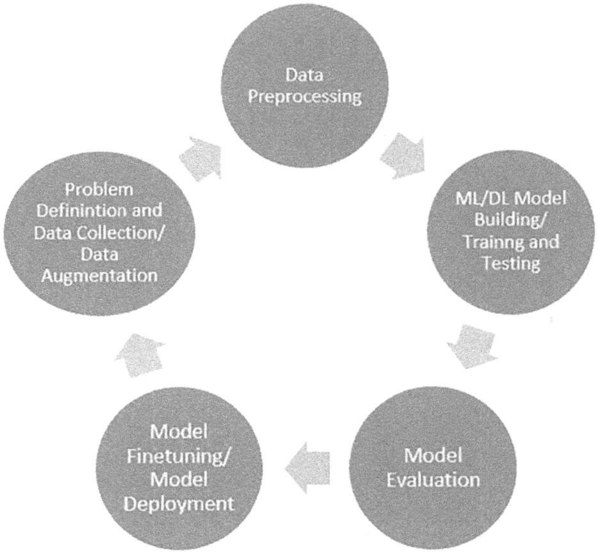

FIGURE 1.1 Generic process workflow of ML/DL model.

cleaning and various feature-engineering tasks, the dataset becomes usable for the model to provide better prediction results.

The next task is to apply the ML/DL models on this preprocessed dataset using the principle of 80–20–20, which means 80% is for training, 20% is for testing and among the training data, 20% is taken for validation.

Based on the evaluation results of the prediction outcome of the model, the parameters will be fine-tuned to get the model fit for deployment. The fundamental metrics applied for this evaluation are Accuracy, Precision, Recall, and F1 score.

The next task is to compare the accuracy defined during the problem definition phase with the model's accuracy. If the accuracy meets the requirement, then stop the model training process and deploy the model. Otherwise, it is essential to fine-tune the hyper-parameters of the model or redefine either the problem definition or dataset regeneration and repeat the whole cycle.

The ML/DL algorithms fall into the two major categories namely Supervised and Unsupervised. Although other types such as Semi-Supervised combine the aspects of both of these, the major focus given in this chapter is on the most important types such as Supervised and Unsupervised learning algorithms.

1.2.1 Supervised vs. Unsupervised Learning

For a learning algorithm, if the dataset has labeled outcomes with class names, then it is called supervised learning. Here, the learning algorithm uses these existing class labels as the basis for predicting the appropriate class of a new data pattern.

Here, a large space of possible hypotheses defined based on the search space (dataset) provided as a hypothesis set acts as the basis for the learning algorithm.

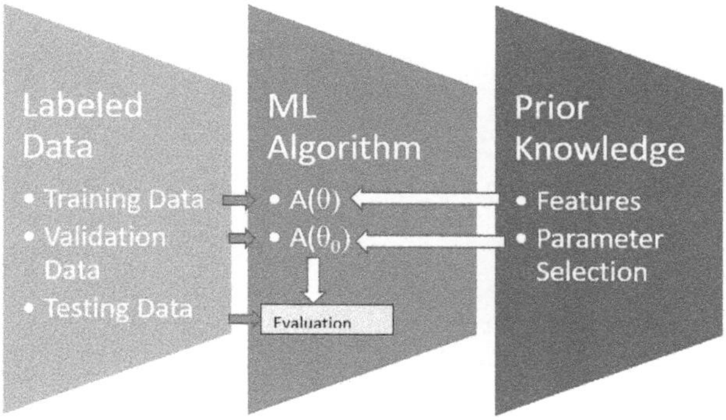

FIGURE 1.2 Supervised learning process flow.

Then the model uses this search space along with its prior knowledge on the same pattern to determine the hypothesis that best fits the newly observed data along with previous knowledge on the same pattern. The supervised learning workflow is given in Figure 1.2.

Some of the ML algorithms come under supervised learning are briefly examined here for basic understanding to the learners as the objective of this book is Explainable AI (XAI).

1.2.2 Linear Regression

The Linear Regression algorithm is a statistical regression method that works on regression to show the relationship between continuous variables. It generally shows the relationship between independent variables plotted in X-Axis and the Dependent variables plotted in Y-Axis. If the number of independent variables is one, then it is simple regression model otherwise it is called multiple linear regression model.

The mathematical equation for Linear regression model is given in Equation (1.1).

$$Y = aX + b \tag{1.1}$$

where Y – outcome/target/dependent variable; X – input/predictor/independent variables; a, b – linear coefficients.

A sample plot for "Boston housing" dataset is given in Figure 1.3. The code used for generating the plot is given below:

```
import numpy as np
import pandas as pd
import matplotlib.pyplot as plt
from sklearn.datasets import fetch_openml
from sklearn.linear_model import LinearRegression
from sklearn.model_selection import train_test_split
# Load Boston Housing dataset from OpenML
boston = fetch_openml(name='boston', version=1, as_frame=True)
```

```
# Features and target
X = boston.data[['RM']] # Using average number of rooms per
  dwelling
y = boston.target.astype(float)
# Train-test split
X_train, X_test, y_train, y_test = train_test_split(
X, y, test_size=0.2, random_state=42
)
# Linear Regression model
model = LinearRegression()
model.fit(X_train, y_train)
# Predictions
y_pred = model.predict(X_test)
# Plot
plt.figure()
plt.scatter(X_test, y_test)
plt.plot(X_test, y_pred)
plt.xlabel("Average number of rooms (RM)")
plt.ylabel("House Price (MEDV)")
plt.title("Linear Regression on Boston Housing Dataset")
plt.show()
```

FIGURE 1.3 Simple linear regression plot for "Boston housing" dataset.

1.2.3 Logistic Regression

Logistic regression is also a regression model that is particularly employed in binary classification. It applies a logit function generally a sigmoid function to find the outcome probability based on the input/predictor/independent variables. The objective of sigmoid function is to convert the output values into probabilities. Based on the threshold value, the probability values are classified into binary class labels. The equation for Logistic Regression is given in Equation (1.2).

$$f(x) = 1/\left(1+e^{-x}\right) \tag{1.2}$$

Here, $f(x)$ – probability outcome; the sigmoid function used here is given in the right hand side.

In Logistic Regression model, based on the outcome, the parameter weights are adjusted so that the model can be trained appropriately till the decision point is found. If the probability value > threshold, then the decision will belong to one class; otherwise, it will belong to the other class.

1.2.4 Decision Tree

Decision trees can be used to model complex nonlinear relationships. They can be used for both nonlinear monotonic and nonlinear nonmonotonic relationships. Here, nonmonotonic means a change in the input could result in a change in the output in any direction and at a varying rate. Monotonic means a change in the input results in a change in the output in the same direction.

Decision tree model is used for both classification and regression. The methodology of decision tree is to recursively split the given data based on "Information Gain," a value calculated from "Gini Index" or "Entropy."

$$\text{Gini index} = 1 - \Sigma(pi)^2 \tag{1.3}$$

$$\text{Entropy}(s) = -\Sigma\left(pi\log_2(pi)\right) \tag{1.4}$$

By applying a recursive cut, the tree-like structure will be created and is used to make a decision. The internal nodes in a decision tree contribute a decision and each outcome is taken from the leaf node in that branch.

1.2.5 Random Forest

Supervised ML technique Random Forest is popular for regression and classification problems. A Random Forest algorithm's accuracy and problem-solving ability increase with the number of trees. Random Forest is a classifier that averages many decision trees on different dataset subsets to increase predicted accuracy. Ensemble learning uses several classifiers to handle complicated problems and improve model performance. They use the same equations as decision tree to find the split position for each tree. Random forest uses a combination of bagging and multiple decision trees to make predictions. Bagging is an ensemble method where multiple models are trained with different parts of the data. Parts of the data can be repetitive but each model receives different datasets to be trained. In random forest each decision tree is trained using different data thereby increasing the robustness of the model. The final result of random forest is the voted result of all the models. This ensures there is very little overfitting while training.

Random forest and other ensemble methods are generally paired up with grid search for hyper-parameter tuning. Grid search takes in the different parameters of a model which can be fine-tuned and runs them exhaustively. The best parameter values are stored. The range and categories of each parameter value can be defined by the user.

1.2.6 K-Nearest Neighbors

KNN stands for K-Nearest Neighbor which is a supervised ML algorithm used for both classification and regression. It is nonparametric, that is the algorithm does not have any parameters to build during the training of data. It uses the existing dataset and makes predictions by comparing the new instances to the trained dataset. It relies on distance metrics such as Euclidean distance, Manhattan distance, etc. to quantify the similarity or dissimilarity between data points.

$$(x, y) = \sqrt{}(x1 - x2) + (y1 - y2)\wedge 2) \quad (1.5)$$

$$(x, y) = |x\_1 - x\_2| + |y\_1 - y\_2| \quad (1.6)$$

Equations (1.5) and (1.6) represent the functions used to find the distance to find the nearest neighbor. Equation (1.5) is used for finding the Euclidean distance while Equation (1.6) is used in finding Manhattan distance. KNN algorithm uses one of these distance calculation methods to find the distance of the given value with other values present. The class label is assigned based on the highest number of nearest points to the given value.

1.2.7 XGBoost

XGBoost acronym for Extreme Gradient Boosting is mainly applied for both classification and regression. It is basically an ensemble learning technique that considers a combination of several learners to provide a robust predictive model. XGBoost helps to provide efficient and scalable predictions.

$$Y(T) = \Sigma(yi, y\_pred, i) + \Sigma\Omega(f) \quad (1.7)$$

The objective function of XGBoost is given in Equation (1.7).

There are two ways by which the performance of the model will be improved. The first one is boosting and the other one is bagging.

XGBoost uses the loss value from the computed result to improve the prediction result so that the performance of the model will be boosted. Here, the boosting technique makes use of the previous iteration result too for training so that the performance will be boosted.

Whereas, the bagging technique uses traditional, independent way of parallel training on each of the models in the ensemble model. It applies both row sampling and column sampling approaches.

1.2.8 Artificial Neural Networks

Biological neural networks formulate the basis for Artificial Neural Networks (ANNs) that mimic the structure of a human brain. It has neurons interconnected with each other and also with neurons in other layers too. In representing this ANN, each node represents one neuron. Each layer can contain any number of neurons which are connected with neurons in the other layers to get the outcome from the neural network. It is shown in Figure 1.4.

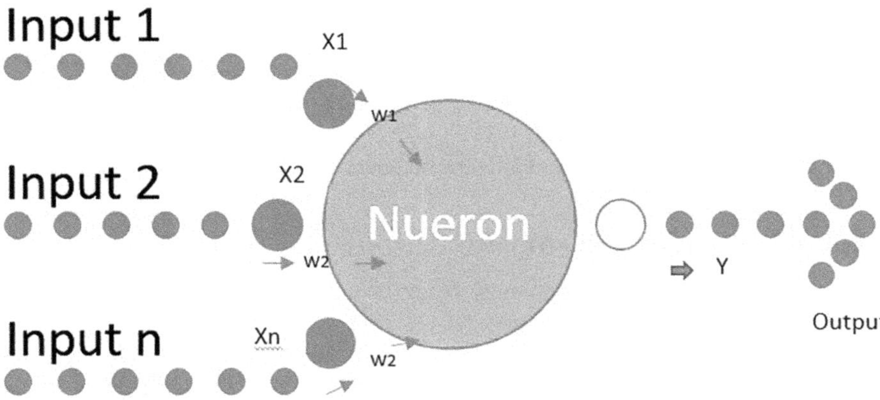

FIGURE 1.4 Artificial neural network.

The ANN technique mimics brain operations. It can be considered as a simple logic gate that gets inputs and applies the OR gate to get the output. The OR gate gives 1 if either of the inputs is true and gives 0 if both are false. Here the output depends on the input passed. In order to mimic the human brain, the error will be calculated from the expected and actual outputs.

If they are same, then we can close the calculation. However, if it is not, then the error will be fed back to the network to update the weight values in order to reduce the bias or error rate.

Any ANN will contain a number of layers. These layers are input layer, output layer and a number of hidden layers. It is shown in Figure 1.5.

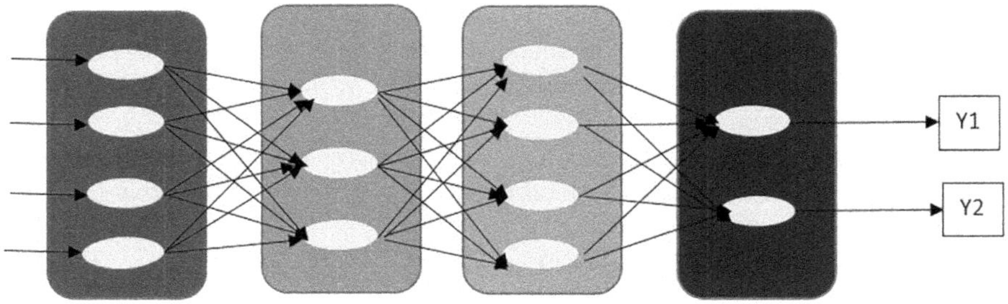

FIGURE 1.5 Simple ANN model with layers.

The input layer gets the inputs from the system or from the user.

The hidden layer performs the transformation operations to convert input to output. All the calculations and processing are done in the hidden layers.

The output layer shows the output from the neural network.

The ANN then computes the weighted sum of the inputs and adds a bias value. This is represented by the following equation.

$$\sum wi \times xi + b \tag{1.8}$$

There is an activation function used to produce the output. These functions are selected based on the application.

For preliminary understanding of DL models, some of the models are discussed below:

1.2.9 Deep Learning (DL) Models

Deep Learning models are completely black boxes as their internal workings are completely unknown to the end users.

As several of the ML models such as Linear Regression, Decision Tree, and GAMs are intrinsically interpretable, they need only some levels of detail in the explanation. However, the working of neural networks and deep neural network models are black boxes in nature, it is essential to provide explanations by means of separate XAI models. These models will help to get the interpretations and explanations both locally as well as globally.

However, as these ML and DL models are generally providing output based on the application but do not provide any information on how that decision is being taken, the researchers have analyzed these critical issues so that they can extract visualizations and explanations from the model by means of interpretability and explainability.

1.3 NEED FOR XAI

As per General Data Protection Regulation (GDPR) law, "the right of explanation for all individuals to obtain meaningful explanations of the logic involved when, automated decision making takes place."

Advanced AI models developed for specific use cases such as healthcare, finance, education, insurance, manufacturing, etc. are highly complex and are not interpretable.

As the AI models make "black-box" forecasts without explaining how they did it, the requirement for researchers, organizations, and regulators to comprehend how AI models make recommendations, forecasts, etc. led to "XAI."

The strategies used by a model to interpret its findings in a way that people can comprehend are referred to as AI explainability. XAI helps in justifying why a model has made a particular choice; makes the nonexperts to understand the reasons behind a particular decision. Due to the transparency derived from the explanations, anyone can understand the internal working of a model.

Assume that you are involved in the development of an image processing application that shows cancer region detection from the MRI images. You have applied some DL models and the results indicate the affected region. However, by showing this alone will never help the medical practitioners to take steps like surgery or chemotherapy. It is important to indicate how that region is detected as the affected region, which features are playing the role in detecting it, and whether the region of selection is correct or not. It can't be said that the entire operation is a black box, and the system takes the decision based on the set of rules given in it. This can be visualized as in Figure 1.6.

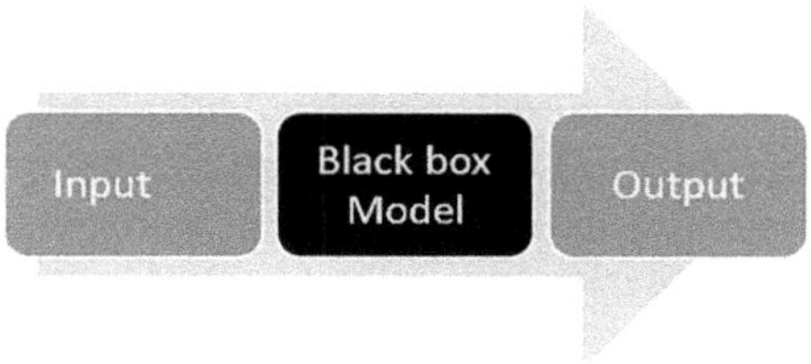

FIGURE 1.6 Black-box model.

Consider another example of designing a predictive analytics model to find the future trends in the fashion industry. It may give some prediction result; however, without knowing on what basis the particular prediction is given, no one can trust the system for deployment. It will be explainable once you dig into the data and features behind the generated results. To get the impact of each of the features on the output is the primary objective of XAI.

Due to the black-box nature of DL models, it is essential to provide explanations on the critical decisions provided by the DL models to ensure ethics, safety and regulatory norms of the application in order to get transparency in the process.

However, it is really challenging to understand how and why the model has taken such a decision. Hence, the explanations derived from the XAI methods must be aligned with the human results and their expectations.

XAI aims in providing explanations of predictions; exploring overall strengths and weaknesses of the model and how it will behave in the future.

XAI is applied in almost all of the smart city and complex real-time applications. Some of them are listed below:

- Smart Healthcare
- Smart Authentication Systems

- Face Recognition System
- Iris Recognition System
- Smart Recommendation Systems
- Smart Cities
- Smart Agriculture
- Smart Finance
- Smart Transportation

1.4 BASICS OF XAI

As per DARPA's XAI program manual, "the stated goal of XAI was to create a suite of new or modified machine learning techniques, enable the end users to understand, appropriately trust, and effectively manage the emerging generation of AI systems."

Some of the essential characteristics that are necessary for the implementation of XAI methods are interpretability, accuracy, and fidelity.

1.4.1 Characteristics of XAI

Interpretability – It refers to the degree to which a model and its predictions can be understood by humans. It is generally related with the model's complexity. Simpler ML models such as Linear Regression, Decision Tree, and Random Forest are inherently interpretable. However, complex models need posthoc explanations to get higher interpretability.

Accuracy – It refers to the model's ability to correctly predict the outcomes on new and unseen instances. This metric helps to decide whether a model can be acceptable for deployment or not.

Fidelity – It is a metric that assesses how well the interpretable model replicates the behavior of a black-box model. The derived interpretations should give simplified explanations of the model's faithfulness to the original model's logic. It is generally evaluated by comparing the interpretable model's predictions with those of the black-box model.

XAI integrates a set of methods and processes that assure the delivery of trustable results generated by the ML/DL algorithms.

As per DARPA, XAI has to generate deep explanations, models must be interpretable and model induction has to be done based on the needs of the application domain.

XAI focuses on the development of interpretable ML/DL systems by addressing challenges in two areas:

1. to develop ML models to handle multimodal data and to develop strategies and ethical policies to identify their working in simulated applications too
2. to develop DL models that can handle complex multimodal datasets so that higher-performance models with explanations meet the ethical policies

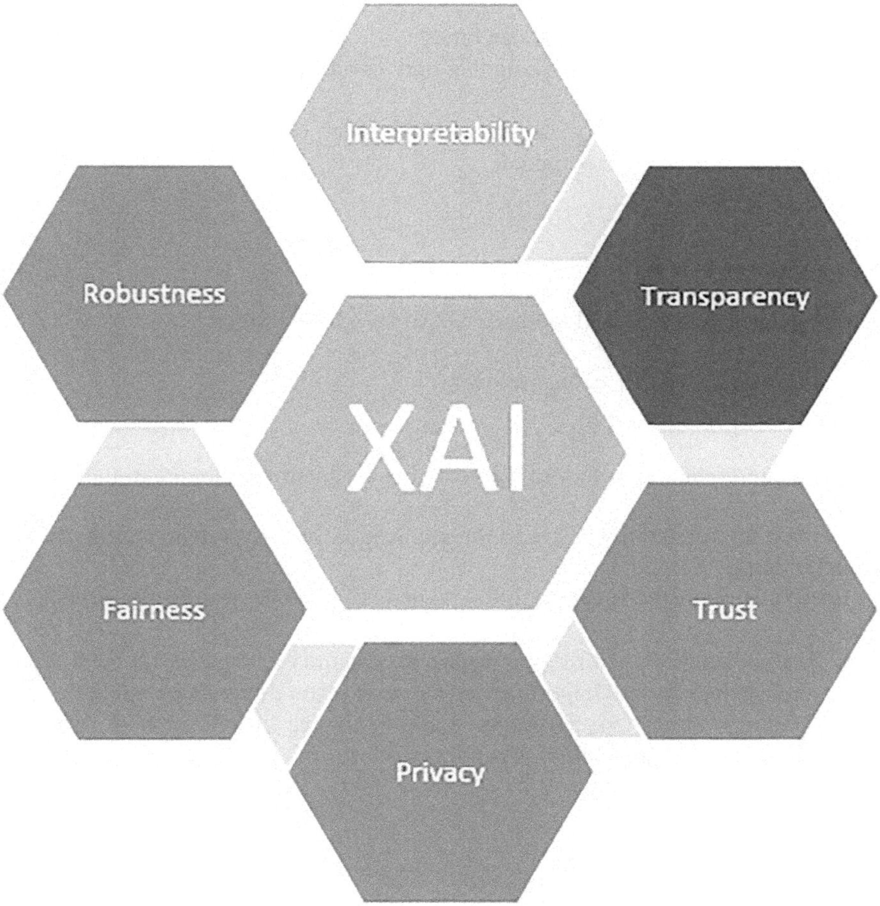

FIGURE 1.7 XAI taxonomy.

1.5 XAI TAXONOMY

The XAI taxonomy is shown in Figure 1.7.

Explaining decisions has the following taxonomy aspects:

- Interpretability: Ease of understanding by the end users on how the decisions are being made.
- Transparency: The model's internal working are transparent as XAI model helps to provide insight on every layer level of the black-box model.
- Fairness: Interpretability helps in ensuring that predictions are unbiased and do not discriminate against underrepresented groups.
- Privacy: It aids in protecting sensitive information in the data.
- Robustness: Interpretability ensures that small changes in the input do not lead to large changes in the prediction.

- Trust: The fidelity measure of XAI helps identify the similarity between the explanations and the actual outcomes.
- Causality: It helps in checking that only causal relationships are picked up by the model.
- Trust: Models that explain their decisions are easier for humans to trust compared to black-box models.

1.6 XAI METHODS

As shown in Figure 1.8, XAI methods can be broadly classified as

- Intrinsically (Pre-hoc) Interpretable
- Posthoc Explainable

Intrinsic interpretability refers to methods that are interpretable due to their simple structure.

Posthoc interpretability refers to the application of interpretation methods after model training.

Interpretation methods can be differentiated based on the results they provide.

- Feature summary statistics → It provides summary statistics for each feature, such as feature importance or pairwise feature interaction strengths.
- Feature summary visualization → allows for the visualization of feature summaries, such as partial dependence plots.

Intrinsic Methods	Posthoc Methods
• Linear/Logistic Regression	• Dimension Reduction (PCA)
• Decision Tree	• Feature Relevance (SHAP/PDP)
• KNN	• Simplified Local/Global Explanation (LIME)
• Rule Based Models	• Visual Explanation (GradCAM, GradCAM++, etc.)
• Bayesian Model	• Text Explanation
	• Attention Mechanisms (GAN)
	• RNN
	• Explanation by example

FIGURE 1.8 Intrinsically interpretable vs. posthoc explainability methods.

- Model internals → such as learned weights or tree structures, can be interpreted for intrinsically interpretable models.
- Data points → such as counterfactual explanations or prototypes, can be used to make a model interpretable.
- Algorithm transparency → provides knowledge of how the algorithm learns a model from the data and what kind of relationships it can learn.
- Intrinsic interpretability:
  - It generally refers to the models that have interpretability as part of the model's working itself. These have simple structures such as decision trees or linear models.
- Posthoc interpretability:
  - It refers to the interpretations derived after model training is completed. Some examples include permutation feature importance or LIME.
- Intrinsic interpretability is limited to simple models.
- Posthoc interpretability can be applied to any model.
- Posthoc interpretability methods can also be applied to intrinsically interpretable models to provide additional insights.

1.6.1 Intrinsically Interpretable Models (Basic ML Models)

Simple ML algorithms have tracing of causal relationships by learning more structured representations of the dataset. As they are intrinsically interpretable, there is no need for additional explanation methods. Interpretable Machine Learning (iML) is the area of research concerned with the creation of interpretable AI systems (interpretable models).

- Interpretability is not required if the model has
  - no significant impact
  - low-risk
  - noncritical applications
  - model's predictions do not have serious consequences
- Model-specific interpretation tools are
  - limited to specific model classes.
  - Example: Interpretation of regression weights in a linear model is a model-specific interpretation.
  - Tools designed for the interpretation of specific models, such as neural networks, are also model-specific.
- Model-agnostic tools can be
  - used on any ML model and are applied after the model has been trained.
  - Example: Agnostic methods analyze feature input and output pairs and do not have access to model interation.
- The interpretation method's scope can be categorized as
  - local,
  - global, or
  - somewhere in between.

- Local interpretation
  - explains an individual prediction, providing insights into why a specific prediction was made.
- Global interpretation
  - aims to explain the entire model behavior, providing an understanding of how the model makes decisions based on a holistic view of its features and learned components.
  - The scope criterion determines whether the interpretation method focuses on individual predictions, the overall model behavior, or a combination of both.
- Algorithm Transparency
  - It shows the internal working of the algorithm on how it learns from the given dataset and finds the relationship patterns to generate the prediction.
  - It shows a better understanding of the working of the ML/DL algorithm but less focus is on the predictions.
- Global, Holistic Model Interpretability
  - Aims to comprehend the entire model at once.
  - Requires knowledge of the trained model, algorithm, and data.
  - Focuses on understanding how the model makes decisions based on a holistic view of its features and learned components.

1.6.2 Properties of XAI Methods

- Expressive Power
  - Describes the "language" or structure of the explanations generated by the method.
  - Examples include IF-THEN rules, decision trees, weighted sums, or natural language.
- Translucency
  - Measures how much the explanation method relies on inspecting the ML model.
  - Ranges from high translucency (relying on model parameters) to zero translucency (only manipulating inputs).
- Portability
  - Defines the range of ML models compatible with the explanation method.
  - Methods with low translucency have higher portability due to treating the model as a black box.
- Accuracy
  - Evaluates how well an explanation predicts unseen data.
  - High accuracy is crucial, especially if the explanation is used for predictions in place of the ML model.
- Fidelity
  - Assesses how well the explanation approximates the prediction of the black-box model.

- High fidelity is essential for the usefulness of an explanation.
- Consistency
  - Examines the variation of explanations between models trained on the same task and producing similar predictions.
  - Consistency is desirable when models rely on similar relationships.
- Stability
  - Focuses on the similarity of explanations for similar instances within a fixed model.
- Comprehensibility
  - Evaluates human understanding of explanations.
  - Considers the audience, size of the explanation, and the comprehensibility of the features used.
- Certainty
  - Addresses the reflection of the model's certainty in the explanation.
  - Includes the model's confidence in predictions and the certainty of different instances.
- Degree of Importance
  - Reflects the importance of features or parts of the explanation.
  - Aims to clarify which conditions or features are most influential in the explanation.
- Novelty
  - Reflects whether a data instance comes from a "new" region.
  - Considers the accuracy and usefulness of explanations for instances outside the training data distribution.
- Representativeness
  - Considers the coverage of explanations.
  - Can cover the entire model or represent individual predictions, impacting the breadth of interpretability.
- Diversity
  - Evaluates the diversity of explanations generated by the method.
  - Aims to provide multiple explanations for the same prediction to increase user trust and understanding.
- Actionability
  - Measures the usefulness of the explanation for decision-making.
  - Considers whether the explanation provides actionable insights for the user.

1.6.3 Characteristics of Good Explanations

- Explanations are Contrastive
  - Discusses the importance of contrasting the current situation with a situation in which the event would not have occurred.
  - Interpretation for ML: Explanations should highlight the difference between the predicted outcome and the alternative outcome to provide a clear understanding of the model's behavior.

- Explanations are Selected
  - Explores the idea that explanations are selective and focus on one or two relevant causes.
  - Interpretation for ML: Explanations should prioritize the most relevant features or factors that contribute to the model's predictions to increase interpretability.
- Explanations are Social
  - Discusses the social nature of explanations and the importance of considering the audience and context.
  - Interpretation for ML: Explanations should be tailored to the user's background, knowledge, and experience to ensure maximum understanding and engagement.
- Explanations Focus on the Abnormal
  - Explores the idea that good explanations focus on abnormal or unexpected events.
  - Interpretation for ML: Explanations should highlight the unusual or unexpected features or factors that contribute to the model's predictions to provide a clear understanding of the model's behavior.
- Explanations are Truthful
  - Discusses the importance of providing truthful and accurate explanations.
  - Interpretation for ML: Explanations should reflect the actual behavior of the model and provide evidence and justification for the model's predictions.
- Explanations are Consistent with Prior Beliefs
  - Explores the idea that good explanations are consistent with prior beliefs and expectations.
  - Interpretation for ML: Explanations should align with the user's prior knowledge and expectations to increase understanding and engagement.
- Explanations are General and Probable
  - Discusses the importance of providing general and probable explanations.
  - Interpretation for ML: Explanations should provide a general understanding of the model's behavior and highlight the most probable factors or features that contribute to the model's predictions.

1.7 APPLICATIONS OF XAI

Kobayashi and Alam (2024) have discussed the use of XAI techniques by incorporating them with their proposed framework consisting of ML classifiers. They have also focused on creating explainable predictable algorithms in their work.

The research work by Anisha Palkar et al. (2024) provides a comprehensive literature survey on the application of ML and XAI in glioma prognosis, highlighting significant advancements in the field [9]. The study underscores the effectiveness of integrating radiomic features from multimodal MR image fusion with ML techniques, as seen in previous research where random forests and SVM models achieved high classification accuracies in distinguishing between different grades of gliomas. A major advantage of this survey is its detailed analysis of feature selection methods and the use of XAI algorithms like Eli5, SHAP, Qlattice, and LIME to enhance the interpretability of ML models, addressing the critical "black-box" issue. However, a limitation identified in the existing literature is the reliance on selected datasets, such as the MIC-CAI BraTS 2019, which may not fully capture the variability in glioma presentations across diverse patient populations.

Have shown prominent results in being able to predict whether a patient is more likely or less likely to suffer from a cardiovascular disease. It was able to achieve accuracies of over 85% on all the different models used and tested. They have used 5771 patients to collect the data. The model developed in this paper uses ECG data primarily to predict cardiac arrests, which it was able to do with a confidence interval of 95%.

Nova (2023) focuses on generative AI such as GANs and VAEs being applied in healthcare for tasks like speech recognition, summarizing medical language, and personalized patient care. The process involves components like speech recognition, natural language processing, and text summarization, and can be integrated with EHR systems. Generative AI can simplify complex medical language into patient-friendly summaries.

Wang et al. (2025) proposes a new approach to personalized content generation based on AI and user instructions. The goal was to overcome the limitations of conventional retrieval-based recommender systems through the automated creation or repurposing of items to cater to the varying needs of users.

Shokrollahi et al. (2023) Denoising Diffusion Probabilistic Models (DDPMs) and Transformer models have been integrated into recent advancements in healthcare applications. Transformer models have been used in many healthcare domains, including protein structure prediction, clinical documentation, and diagnostic assistance.

Zhang and Kamel Boulos (2023) address how Generative Pretrained Transformer (GPT) models are developing quickly and how they can revolutionize healthcare, especially in the areas of prognosis, diagnosis, and clinical decision support.

Cao et al. (2025) in their paper offered a thorough examination of the development, history, and uses of generative AI models. It divides the models into multimodal and unimodal categories, emphasizing notable advancements in AI-generated content tasks in various industries. They have highlighted the widespread use of encoder-decoder models and transformer architecture, including the Text-to-Text Transfer Transformer and Switch Transformer. They also discuss how generative models are used in computer vision, natural language processing, and image creation.

Ivanovic et al. (2023) discuss the influence of AI techniques in processing medical big data and their applications in contemporary society. They highlight the need for extensive patient data for the development of intelligent decision support systems. The use of AI techniques in modeling and simulation of drug development is also explored, emphasizing the potential for accelerating effective new drug finding. Additionally, the authors present various AI applications in healthcare, including personalized predictions, treatments, and recommendations, as well as the use of AI in information-seeking dialogues and autonomous agents in healthcare.

The review given in (2023) focuses on the use of DL, particularly CNNs, in detecting and classifying brain strokes via medical imaging. The paper highlights the effectiveness of CNNs in classifying strokes into normal, hemorrhagic, and ischemic categories, with a focus on the enhanced accuracy provided by transfer learning and hybrid models. The study finds that DL models, especially CNNs like ResNet152, which achieved 99.25% accuracy, are becoming increasingly crucial in stroke diagnosis. These advancements improve diagnostic speed and precision, offering significant potential for better patient outcomes through early detection and treatment.

Saraswat et al. (2022) conducted a survey of the various applications of XAI in different fields. The paper shifted focus toward the use of DL techniques and XAI in healthcare. It also highlighted the use of XAI for dimensionality reduction in healthcare applications.

Further, Chae et al. (2022) have implemented an early warning system to alert the patients who are likely to suffer from a cardiac arrest. This was done using ML and DL algorithms such as decision tree, random forest, LSTM, LSTM-GRU hybrid model. The LSMT model proposed was able to predict results with an accuracy of 85.92%.

Insights can be taken from the study conducted by Yang (2022) which highlights the significance of XAI in healthcare systems. It highlights the need for XAI and how it can remove the barrier of trust in AI models especially in critical scenarios such as healthcare systems.

Loh et al. (2022) provide insights on different XAI techniques such as GradCAM, LIME, Fuzzy classifier and SHAP used along with various ML and DL models in the context of COVID-19 detection and cancer classification.

The authors in (2022) propose a novel deep learning system (DLS) that integrates four models – blood supply, border zone, brain structure atlas, and cortex – to provide a multidimensional approach for classifying and subtyping ischemic stroke. The system optimizes image preprocessing and rearranges CNN output channels to improve convergence speed and accuracy in locating infarction lesions using MRI-DWI images. Validated by two experienced neuroradiologists, approximately 90% of the system's results were deemed clinically acceptable. The study highlights the DLS's potential in automatic ischemic stroke classification and subtyping, demonstrating reliability across a dataset of 5,636 samples and independent tests. Future work will focus on optimizing the DLS for varying imaging conditions across different devices and institutions.

Authors (2022) have explored the use of DL models, particularly VGG-16, in medical imaging for ischemic stroke detection. The paper highlights how DL has advanced disease classification, with VGG-16 showing effectiveness in various applications, including diabetic retinopathy, brain tumor classification, and COVID-19 detection. The study finds that VGG-16 achieved a high validation accuracy of around 90% for ischemic stroke detection, outperforming other models like ResNet50 and InceptionV3. This high accuracy underscores VGG-16's ability to capture detailed features in medical images. However, the paper notes limitations such as dataset size, diversity, and reliance on a single imaging modality, which could affect the model's generalizability. Future research should focus on expanding datasets, incorporating diverse imaging modalities, and integrating XAI techniques to enhance model robustness and trust in clinical settings.

XAI is critically reviewed by Angelov et al. (2021) within the framework of recent developments in DL and ML. It outlines the difficulties in reaching XAI and talks about different approaches and potential future research areas. They highlighted the significance of XAI for regulatory purposes. Furthermore, they investigated how prototype-based models might be used in the context of DL to produce a representation that is easily understood.

Zhao et al. (2026) highlight different techniques such as content-based, hybrid, and collaborative filtering-based approaches and their use in healthcare. They describe five use cases for health recommendation systems, including disease prediction, patient and physician decision-making, training, diet, and lifestyle recommendations.

Briefly discuss the use of AI techniques to improve patient health and address quality-of-life issues in the medical and healthcare domains. They explore the use of AI techniques in processing medical big data and providing personalized services, as well as predictive analytics in the medical fields.

Based on what authors Dave et al. (2024) mentioned, we understand the difference between the use of LIME and SHAP. As highlighted in the paper LIME can provide local explanations while SHAP can help in providing global explanations. The explainable models are deployed with XGBoost classification model.

The importance of clinicians comprehending the underlying reasoning behind AI predictions is emphasized by Lauritsen et al. (2020) who addressed the need for transparency and explainability in AI models for clinical practice. The XAI-EWS (XAI early warning system), an AI model created to forecast acute critical illness from electronic health records and offer clear, visual explanations for those predictions, is introduced in this article. The model makes use of a Deep Taylor Decomposition (DTD) explanation module in addition to a Temporal Convolutional Network (TCN) for predictions.

Suryadevara (2026) discusses the development and evaluation of an intelligent medication recommendation system that uses AI and ML to provide personalized medication recommendations. It collects patient data, preprocesses it, and uses statistical techniques for feature extraction to analyze patient profiles and medical histories.

Sharma et al. (2020) focused on the trade-off between accuracy and explainability. It elaborates on the importance of AI in the medical domain while providing necessary importance on how it should be transparent and prioritize safety.

The authors Wiens and Shenoy (2018) talk about the use of ML in processing electronic health records. While being able to provide risk analysis, the paper mentions that separate statistical analysis will be required to be able to use ML in the healthcare sector.

Wiens and Shenoy (2018) provided an in-depth overview of medical image segmentation techniques, with a focus on mammogram tumor segmentation. The paper surveys previous research utilizing the 'INBreast' dataset, commonly used in breast cancer studies, particularly for mass segmentation and other related tasks. The review highlights the importance of segmentation in medical imaging, crucial for identifying anatomical structures and disease- specific features. It discusses advancements like dilated convolution, which expands the receptive field of CNNs without sacrificing spatial resolution, essential for detecting subtle details in images. The authors also explore semi-dilated convolution as a method to improve segmentation accuracy while preserving resolution. The paper addresses challenges like pixel-level discrepancies and the need for model interpretability, which are key to developing effective AI-assisted diagnosis systems.

1.8 ROLE OF XAI IN ETHICAL AI SYSTEMS DEVELOPMENT

It is evident that if the training data has biases, naturally the AI algorithms trained on it will have issues in delivering the results. This will lead to ethical concerns as several crucial decisions are based on the model's outcome. AI ethics helps to establish the rules and regulations to give responsible and unbiased results from the AI models.

To achieve this, XAI provides understandable and interpretable explanations in terms of text and visual outcomes for the AI models' predictions. Using these explanations, the model developers and engineers can identify the biases in the model and can fine-tune the model appropriately. XAI promotes accountability by giving insights on how the model behaves under different settings and whether the decisions taken from the model have any biases or ethical breaches. Further, XAI helps to promote fairness by providing trustworthiness in AI systems in terms of their recommendations and evaluations. It is shown in Figure 1.9.

XAI also ensures transparency in the decision-making process by showcasing how each part of the model's functionality plays its role in reaching the decision. For instance, when it is applied to DL models, it can highlight the working of each layer-wise, each neuron-wise visualization.

In addition to that, XAI evaluation metrics such as Fidelity, Unambiguity, and Interpretability size also help in validating the model's trustworthiness and also the best XAI method that can be applied for getting responsible AI deployment.

Some of the techniques applied in XAI such as counterfactual explanations, iML, exposure of biased, and sensitive attributes that affect the outcome, attention

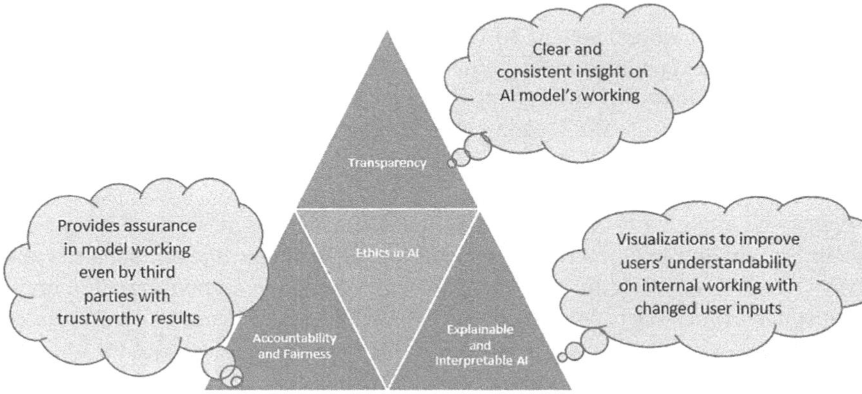

FIGURE 1.9 Ethics in AI using XAI.

mechanisms, and plots for better visualizations help in achieving transparency, fairness, accountability, and trustworthiness of AI models.

XAI helps achieve Ethics in AI in the following ways:

1. Mitigation of Bias – Using XAI's explanations, it will be easier to identify potential biases in the algorithm as well as in the data.
2. Ensuring Fairness – Especially in critical areas such as law, criminal justice, loan processing, recruitment process, etc., the AI models should ensure fairness. Using XAI model explanations, one can ensure that the AI models are fair to use.
3. Fidelity of decisions – By applying XAI methods, the users can trace how the process has taken place in order to reach a particular decision. Using the explanations, people can understand how a particular decision is arrived at and what the trustworthiness of the model's outcome is.
4. Accountability and Auditability – As it is possible to visualize the internal working by changing the user inputs dynamically, XAI helps improve accountability of the decision-making process. If negative consequences are derived, then the AI model can be revised and modified to become a trustworthy one.
5. Compliance with Ethical and Legal Standards – XAI promotes privacy and transparency by means of real-time monitoring in the AI decision-making process. Hence, it is possible to monitor and ensure that whether ethical and legal standards have been followed in AI development.
6. Improved User Understanding – As XAI provides complete visualization and explanations on how the features in the dataset are impacting the output, it provides improved understanding of the internal working of the model.

Hence, it is implied that XAI imparts ethical guidelines and leads to responsible AI model development. It is beneficial to the entire human race as it promotes fair, accountable and transparent insights on the working of AI models.

1.9 POLICY GUIDELINES IN AI SYSTEMS DEVELOPMENT

For AI systems development, policies should give guidelines on how the systems should be developed with trustworthiness using XAI. In order to ensure responsible and ethical AI practices, these guidelines should focus on data usage, development of model, testing and deployment, and continuous monitoring.

The ethical principles as imposed in policy guidelines should emphasize AI usage by individuals and society. It ensures unbiased outcomes for all users without discrimination. AI ethical guidelines ensure protection of data privacy and sensitivity. The ethical policy guidelines outline that every user of the system should be able to understand the model's working and provide reliable access to the systems. These policy guidelines also ensures that error detection and error correction should be addressed without any side effects on the normal working of the system.

Policies with reference to data are related to data quality and the training has to be done on different types of data and benchmark datasets in order to avoid bias; protection of data from unauthorized access and prevention of security breaches.

Other policy guidelines should be established for data protection with regulatory compliance.

In addition to that, XAI techniques should be implemented with the objectives of bias detection and mitigation, reliability testing of models, continuous monitoring of model performance degradation, introduction of new bias due to input data change, audits for ethical guidelines compliance, and feedback mechanisms to get insights from all the stakeholders of AI-based applications.

Generally, AI policy should contain the following major components:

1. Ethical Guidelines for Responsible AI development
2. Implementation of data protection with data privacy laws
3. Guidelines for Transparency and explainability requirements
4. Mitigation measures for potential biases
5. Accountability and governance guideline framework

XAI plays a crucial role in exploring all of the above by providing insights on the internal working of the DL models.

1.10 SUMMARY

This chapter focuses on the introduction part of XAI taxonomy. The AI systems are currently highly complex in nature and their decisions or predictions need to be proved for their trustworthiness. Hence, the researchers have focused on developing methods that can provide explanations and visualizations of the internal working of such "black-box" models. In view of that, this chapter provided the distinction between the intrinsically interpretable and Posthoc explainable models. From the various foundation models for real-time data analytics, the ML models such as regression models and decision tree are self-explanatory in the sense that they are intrinsically interpretable and from the model's outcome and intermediate results one can get insights on how the decision is arrived at. However, when it comes to ensemble techniques, GAN, ANNs, advanced deep neural network models, they are not intrinsically interpretable. Their complex processing units and internal architecture make them completely "black box." Such models are called Posthoc explainable models as they need to be analyzed for the reason of decision-making after the predictions are received. In this regard, this chapter provides insight on the basic differences between interpretability and explainability. Further, the chapter also dealt with the ethics and policy guidelines to be followed whenever such complex AI systems are employed for real-time analytics.

PRACTICE QUESTIONS

I. Short Answers – Remember / Understand
   1. List the characteristics of intelligent systems.
   2. Define Machine Learning.
   3. Differentiate supervised and unsupervised learning.

4. Prepare a list on the need for XAI.
5. Name some of the applications in which explanations are mandatory. Justify.

II. Descriptive Qns. – Apply / Analyze
1. Elaborate the characteristics of XAI with the application of a real-word case study.
2. Showcase the generic workflow of XAI with reference to loan sanction application of a bank.
3. Which models provide global interpretability? How are they different from black-box models?
4. How does an XAI-integrated framework help in an eye disease classification system? Examine its applicability in AI systems development.

III. Higher Order Thinking Questions – Analyze / Create / Evaluate
1. Justify that the decision tree ML model is intrinsically interpretable with a case study application of your own.
2. Assume that you are involved in stable object detection of a video by applying deep neural network. You are now in need of proving that it will work as expected in a real-time scenario. How will you ensure the trustworthiness of the model?
3. Do you think ethics is mandatory in AI systems development? If so, elaborate on how you will ensure it.
4. Describe the application of policy guidelines related to XAI to achieve ethics and responsiveness in AI systems.

BIBLIOGRAPHY

Angelov, P. P., Soares, E. A., Jiang, R., Arnold, N. I., and Atkinson, P. M. (2021). Explainable artificial intelligence: an analytical review. *Wiley Interdisciplinary Reviews: Data Mining and Knowledge Discovery*, 11(5):e1424.

Bahalul Haque, A. K. M., Najmul Islam, A. K. M., and Mikalef, Patric. (2023). Explainable Artificial Intelligence (XAI) from a user perspective: A synthesis of prior literature and problematizing avenues for future research. *Technological Forecasting and Social Change*, 186(Part A):122120. ISSN 0040-1625, https://doi.org/10.1016/j.techfore.2022.122120

Cao, Yihan, Li, Siyu, Liu, Yixin, Yan, Zhiling, Dai, Yutong, Yu, Philip, and Sun, Lichao. (2025, May). A survey of AI-generated content (AIGC). *ACM Computer Survey*, 57(5):Article 125, 38 pages. https://doi.org/10.1145/3704262

Cecilia, Panigutti, Alan, Perotti, and Dino, Pedreschi (2020). Doctor XAI: An ontology-based approach to black-box sequential data classification explanations. In *Proceedings of the 2020 Conference on Fairness, Accountability, and Transparency (FAT'20)*, Barcelona, pp. 425–435.

Cecilia, Panigutti, Andrea, Beretta, Daniele, Fadda, Fosca, Giannotti, Dino, Pedreschi, Alan, Perotti, and Salvatore, Rinzivillo (2022). Co-design of human-centered, Explainable AI for clinical decision support. *ACM Transactions on Interactive Intelligent Systems*, 13(4): 1–35.

Chae, M., Gil, H.-W., Cho, N.-J., and Lee, H. (2022). Machine learning-based cardiac arrest prediction for early warning system. *Mathematics*, 10(12):2049. https://doi.org/10.3390/math10122049

Clement, T., Kemmerzell, N., Abdelaal, M., and Amberg, M. (2023). XAIR: A systematic metareview of Explainable AI (XAI) aligned to the software development process. *Machine Learning and Knowledge Extraction*, 5(1):78–108. https://doi.org/10.3390/make5010006

Dave, V., Borade, H., Agrawal, H., Purohit, A., Padia, N., and Vakharia, V. (2024). Deep learning-enhanced small-sample bearing fault analysis using Q-transform and HOG image features in a GRU-XAI framework. *Machines*, 12(6):373.

Dino, Pedreschi, Fosca, Giannotti, Riccardo, Guidotti, Anna, Monreale, Luca, Pappalardo, Salvatore, Ruggieri, and Franco, Turini. (2018). Open the black box data-driven explanation of black box decision systems. *Arxive preprint arXiv:1806.09936.*

Došilović, F. K., Brčić, M., and Hlupić, N. (2018). Explainable artificial intelligence: A survey. In *Proceedings of the 41st International Convention on Information and Communication Technology, Electronics and Microelectronics (MIPRO)*, Opatija, Croatia, pp. 210–215. https://doi.org/10.23919/MIPRO.2018.8400040.

Farrow, Robert (2023). The possibilities and limits of XAI in education: A socio-technical perspective. *Learning, Media and Technology*, 48(2):266–279. https://doi.org/10.1080/17439884.2023.2185630

Frąckiewicz, Marcin. (2023). Explainable AI (XAI) in education: Enhancing learning and personalized instruction, artificial intelligence. *TS2 Space.*

Ivanovic, M., Autexier, S., Kokkonidis, M., and Rust, J. (2023). Quality medical data management within an open AI architecture – Cancer patients case. *Connection Science*, 35(1). https://doi.org/10.1080/09540091.2023.2194581

Khosravi, Hassan, Shum, Simon Buckingham, Chen, Guanliang, Conati, Cristina, Tsai, Yi-Shan, Kay, Judy, Knight, Simon, Martinez-Maldonado, Roberto, Sadiq, Shazia, and Gašević, Dragan. (2022). Explainable artificial intelligence in education. *Computers and Education: Artificial Intelligence*, 3:100074. ISSN 2666-920X, https://doi.org/10.1016/j.caeai.2022.100074

Kobayashi, Kazuma, and Alam, Syed Bahauddin. (2024). Explainable, interpretable, and trustworthy AI for an intelligent digital twin: A case study on remaining useful life. *Engineering Applications of Artificial Intelligence*, 129. https://doi.org/10.1016/j.engappai.2023.107620

Lauritsen, S. M., Kristensen, M., Olsen, M. V., Larsen, M. S., Lauritsen, K. M., Jørgensen, M. J., Lange, J., and Thiesson, B. (2020). Explainable artificial intelligence model to predict acute critical illness from electronic health records. *Nature Communications*, 11(1):3852.

Loh, Hui Wen, Ooi, Chui Ping, Seoni, Silvia, Barua, Prabal Datta, Molinari, Filippo, and Rajendra Acharya, U. (2022). Application of explainable artificial intelligence for healthcare: A systematic review of the last decade (2011–2022). *Computer Methods and Programs in Biomedicine*, 226. https://doi.org/10.1016/j.cmpb.2022.107161

Mattia, Setzu, Riccardo, Guidotti, Anna, Monreale, Franco, Turini, Dino, Pedreschi, and Fosca, Giannotti. (2021). GLocalX – From local to global explanations of black box AI models. *Artificial Intelligence*, 294: 103457.

Nogueira, Ana Rita, Pugnana, Andrea, Ruggieri, Salvatore, Pedreschi, Dino, and Gama, João. (2022). Methods and tools for causal discovery and causal inference wires data mining and knowledge discovery. *Wires Data Mining and Knowledge Discovery*, 12(2):e1449.

Nova, K. (2023). *Generative AI in Healthcare: Advancements in Electronic Health Records, Facilitating Medical Languages, and Personalized Patient Care.*

Nwakanma, C. I., Ahakonye, L. A. C., Njoku, J. N., Odirichukwu, J. C., Okolie, S. A., Uzondu, C., Ndubuisi Nweke, C. C., and Kim, D.-S.. (2023). Explainable Artificial Intelligence (XAI) for intrusion detection and mitigation in intelligent connected vehicles: A review. *Applied Sciences*, 13(3):1252. https://doi.org/10.3390/app13031252

Orestis, Lampridis, Riccardo, Guidotti, and Salvatore, Ruggieri (2021). Explaining sentiment classification with synthetic exemplars and counter-exemplars, discovery science. In *Proceedings of the International Conference on Discovery Science*. Cham: Springer, pp. 357–373.

Palkar, A., Dias, C. C., Chadaga, K., and Sampathila, N. (2024). Empowering Glioma prognosis with transparent machine learning and interpretative insights using Explainable AI. *IEEE Access*, 12:31697–31718.

Riccardo, Guidotti, Anna, Monreale, Fosca, Giannotti, Dino, Pedreschi, Salvatore, Ruggieri, and Franco, Turini. (2021). Factual and counterfactual explanations for black box decision making. *IEEE Intelligent Systems*, 34(6):14–23.

Riccardo, Guidotti, Anna, Monreale, Salvatore, Ruggieri, Franco, Turini, Fosca, Giannotti, and Dino, Pedreschi. (2022). A survey of methods for explaining black box models. *ACM Computing Surveys (CSUR)*, 51(5):1–42.

Saeed, Waddah, and Omlin, Christian. (2023). Explainable AI (XAI): A systematic meta-survey of current challenges and future opportunities. *Knowledge-Based Systems*, 263:110273. ISSN 0950-7051. https://doi.org/10.1016/j.knosys.2023.110273

Saraswat, D., Bhattacharya, P., Verma, A., Prasad, V. K., Tanwar, S., Sharma, G., Bokoro, P. N., and Sharma, R., (2022). Explainable AI for healthcare 5.0: Opportunities and challenges. *IEEE Access*, 10:84486–84517. https://doi.org/10.1109/ACCESS.2022.3197671.

Sharma, D., Koundilya, V., and Verma, S. (2020). Explainable AI(XAI): A review. *International Journal of Psychosocial Rehabilitation*, 24(5), 56498–56502.

Sharma, B., Sharma, L., Lal, C., and Roy, S. (2024). Explainable artificial intelligence for intrusion detection in IoT networks: A deep learning based approach. *Expert Systems with Applications*, 238:121751.

Shokrollahi, Y., Colmenarez, J., Liu, W., Yarmohammadtoosky, S., Nikahd, M. M., Dong, P., Li, X., and Gu, L. (2023). Recent advances in generative AI for healthcare applications. *Journal of Imaging Informatics in Medicine*. https://doi.org/10.1007/s10278-026-01908-0

Suryadevara, N., Priya, V., and Sharma, S. (2026). From black box to clear box: Explainable AI for next-gen pharmacovigilance. *Expert Opinion on Drug Safety* (just-accepted).

Theissler, Andreas, Spinnato, Francesco, Schlegel, Udo, and Guidotti, Riccardo. (2022). Explainable AI for time series classification: A review, taxonomy and research directions. *IEEE Access*, 10:100700–100724.

Tjoa, Erico, and Guan, Cuntai. (2020). A survey on explainable artificial Intelligence (XAI): Towards Medical XAI. *IEEE Transactions on Neural Networks and Learning Systems*, 32(11):4793–4813.

Vale, Daniel, El-Sharif, Ali, and Ali, Muhammed. (2022). Explainable artificial intelligence (XAI) post-hoc explainability methods: Risks and limitations in non-discrimination law. *AI and Ethics*, 2(4):815–826. https://montrealethics.ai/explainable-artificial-intelligence-xai-post%E2%80%90hoc-explainability-methods-risks-and-limitations-in-non%E2%80%90discrimination-law

Wang, Wenjie, Lin, Xinyu, Feng, Fuli, He, Xiangnan, and Chua, Tat-Seng. (2025). Generative recommendation: Towards personalized multimodal content generation. In *Companion Proceedings of the ACM on Web Conference (WWW'25)*. Association for Computing Machinery, New York, pp. 2421–2425. https://doi.org/10.1145/3701716.3717529

Wiens, J., and Shenoy, E. S. (2018, January 6). Machine learning for healthcare: On the verge of a major shift in healthcare epidemiology. *Clinical Infectious Diseases*, 66(1):149–153. https://doi.org/10.1093/cid/cix731. PMID: 29020316; PMCID: PMC5850539.

Yang, S., Folke, T., and Shafto, P. (2022). A psychological theory of explainability. In *Proceedings of the 39th International Conference on Machine Learning*, Baltimore, MD.

Zhang, P., and Kamel Boulos, M. N. (2023). Generative AI in medicine and healthcare: Promises, opportunities and challenges. *Future Internet*, 15(9):286. https://doi.org/10.3390/fi15090286

Zhao, Y., Ullah, F., Mahmood, K., Saeed, S., Mohammad, N., and Raza, U. (2026, February). XAI-EdgeSFL: explainable edge intelligence with adaptive intrusion-resilient split federated learning for consumer healthcare ecosystems, *IEEE Transactions on Consumer Electronics*, 72(1):2185–2196. https://doi.org/10.1109/TCE.2026.3656201

2 XAI Techniques and Models

2.1 INTRODUCTION TO XAI TECHNIQUES

The explanations derived from XAI techniques and models are usually the visual interpretations based on feature importance scores derived from simpler surrogate models or from focused data point values. XAI techniques can be categorized into four ways. They are:

- Model-Agnostic
- Model-Specific
- Global Explainability
- Local Explainability

Model-agnostic and model-specific methods

- XAI methods can analyze models in two ways: **model-agnostic** or **model-specific**. Model-agnostic methods aren't limited by the characteristics of an AI algorithm and can be used to reveal the decision process of any black-box model. Simply put, they're a generic tool for developers to analyze any model. Model-specific methods, on the other hand, are tailored to analyze specific algorithm's characteristics and provide a more detailed understanding of how a certain algorithm improves model decision-making. However, these methods require a higher degree of skill from developers to analyze the models.
- Different XAI methods have varying perspectives on analyzing models and data. For example, the SHAP values of a model can be used to identify positive and negative impacts of data features on the model outcome – like are the impacts of salary and credit card score in making a decision to grant a loan to an applicant positive or negative, and how big is their significance? Another example is counterfactual explanations, where different cases can be generated that eventually show the inputs with which the model gives a specific result. XAI methods can also be categorized into global or local explanations.

Let us discuss each of these techniques with an example.

DOI: 10.1201/9781003630241-3

2.2 Model-Agnostic

These techniques are usually provided with Machine Learning Models. They are not usually limited by the AI models. They are highly generic in nature and help to analyze the features that are contributing to the model's outcome. Some of the typical model-agnostic techniques are: Local Interpretable Model-Agnostic Explanations (LIME), SHAP, and PDP (Partial Dependency Plots).

2.2.1 LIME

Local Interpretable Model-Agnostic Explanations is a model-agnostic, local explanation technique. It generates explanations for individual predictions based on individual data values, thus providing local explanations rather than providing global interpretations.

It generally provides a simple, linear visualization about a data point from a complex ML model. It can show each feature as a data point and modifies it to find out its impact on the models' output. The input features are perturbed and the resultant changes in the model's output are used to determine which features are influencing the model's outcome. The LIME technique can be applied to any kind of multimodal data such as tabular, image, video, audio, or signal. The generic workflow of LIME is given in Figure 2.1.

The LIME model's working is mathematically given below.

Consider a local data point "x" which needs to be analyzed for its impact on the model's outcome, the LIME equation is given below:

$$E(x) = \arg\min L\left(f, f^{'}, lx\right) + C\left(f^{'}\right), \quad f^{'} \in f$$

where

f is the actual objective function which represents the ground truth.

$f^{'}$ is a surrogate function or explanation function to approximate f in the proximity of x. It belongs to the family of linear functions.

lx defines the locality of reference.

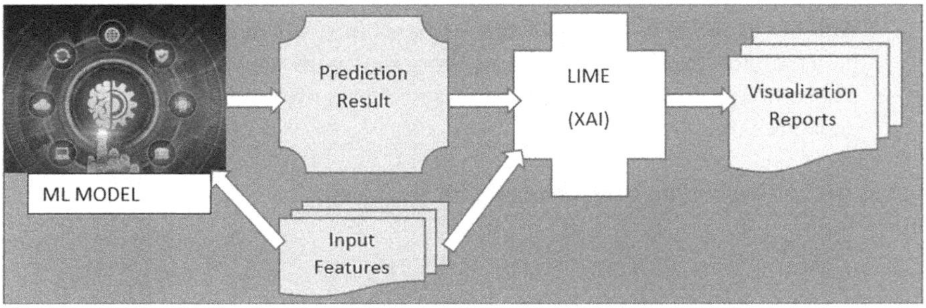

FIGURE 2.1 LIME generic flow.

L is the fidelity function or loss function that needs to be minimized on local mismatch between f and f' and can be visualized as RMSE (R-Mean Square Error).

C is the complexity measure which is opposed to explainability.

Example

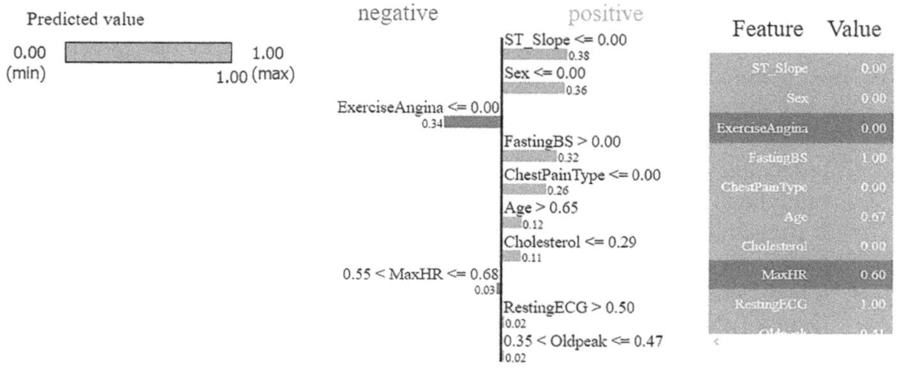

FIGURE 2.2 LIME visualization of Cardiovascular Health Care Data for Logistic Regression Model Prediction on risk of heart attack.

The LIME visualization shown in Figure 2.2 provides the features importance related to the outcome predicted by the Logistic Regression model. It indicates, the model's prediction is positively decided by some features and is negatively impacted by "ExcerciseAngina." The right-most block is the explanation for a particular input data point. From this visualization, one can easily get interpretation that, as the input data has age>67; Fasting BS is high and Resting ECG is high, they positively impact on "Heart Attack Risk."

Advantages of LIME

- LIME is a Model-Agnostic technique.
- It provides Local Interpretability.
- The explanations derived from LIME are Human-Understandable.
- LIME plots provide Feature Importance Ranking.
- It has Visualization Capability.
- LIME has Versatility based on the application and dataset.

Disadvantages of LIME

- LIME visualizations will not be accurate and hence have approximation error.
- Due to dynamic input data, LIME has instability in its explanations.
- It is computationally expensive.
- As it is a local interpretable technique, it has local focus.
- It is sensitive to perturbation strategy applied to generate surrogate models.
- LIME has issues in handling correlated features as perturbing one data point may affect the correlated feature too.

2.2.2 SHAP

SHAP is also a model-agnostic technique. It is based on game theoretic approach to explain the prediction results of ML models. Unlike LIME, the SHAP technique provides reliable and consistent model predictions.

It provides the visualizations and explanations based on Shapley values. These values are used to find the features' importance on the model's output. This technique applies a random masking method on the features and finds the impact of these masks/changes in the model's output. SHAP model's outcome is very much similar to feature importance plot. It represents both positive and negative impacts of each of the input features on the model's prediction outcome.

The key features of SHAP are listed below:

- For Shapley values calculation, it applies a kernel-based estimation approach using local surrogate models.
- It uses Shapley Values for each feature to find its impact on the model's prediction outcome.
- It uses Additive Feature Attribution method.
- Unlike LIME, even though SHAP is a Model-Agnostic technique, it has both Global and Local Interpretability.
- In SHAP, the prediction is explained by assuming that each feature value of the instance is a "player" in a game. Each prediction task is taken as a game. The gain is nothing but the difference between actual prediction result for an instance or data point and the average prediction of all instances or data points. Each feature is a player in this game. Hence, the gain value calculated will show how impactful a feature is on the model's prediction outcome.

$$\text{Gain} = \text{Actual prediction result for an instance or data point} \\ - \text{Average prediction of all instances or data points}$$

Using SHAP, one can generate two different plots

- SHAP Feature Importance Plot
- SHAP Summary Plot

2.2.2.1 SHAP Feature Importance Plot

SHAP feature importance measured as the mean absolute Shapley values. Each feature's importance on the model's prediction is calculated as above using Shapley values. Here, an application to find the human activity recognition using various features is predicted using CNN model. The visualization on how model has provided the outcome is analyzed using SHAP. The SHAP technique has generated the SHAP feature importance plot as shown in Figure 2.3. It shows the feature 'fBodyAccJerk-bandsEnergy().1.8' has the highest impact on the CNN model's outcome.

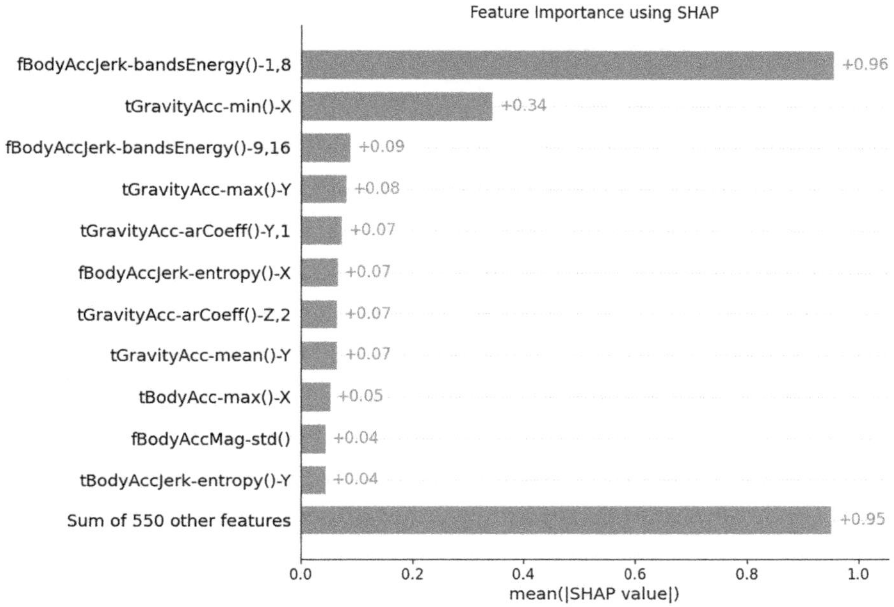

FIGURE 2.3 SHAP feature importance graph.

2.2.2.2 SHAP Summary Plot

For the same application of human activity recognition, a SHAP summary plot is drawn as given in Figure 2.4. It shows the feature 'fBodyAccJerk-bandsEnergy().1.8' has both positive and negative impact and has the highest impact on the model's prediction result.

Advantages of SHAP

- Theoretically Sound
- Consistent and Accurate
- Model-Agnostic
- Local and Global Interpretability
- Fair Attribution
- Handling Feature Interactions

Disadvantages

- Computationally Intensive
- Complexity
- Approximation
- Difficulty with Correlated Features

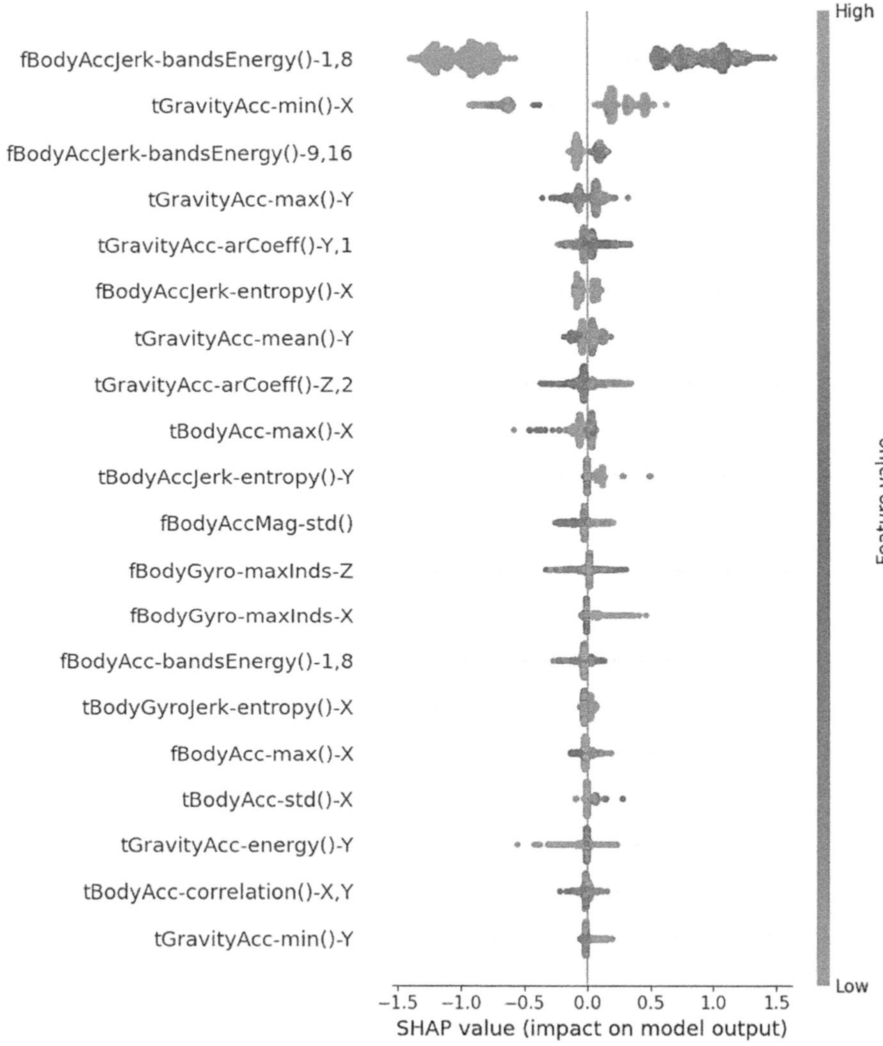

FIGURE 2.4 SHAP summary plot.

2.2.3 TreeSHAP

- Decision trees, random forests, gradient boosting machines (GBMs), and XGBoost come under the TreeSHAP visualization technique. An example of TreeSHAP visualization using Decision Tree ML model is given in Figure 2.5.
- It leverages the structure of these tree-based models to significantly speed up the calculation of SHAP values.

- TreeSHAP provides an optimized algorithm that:
  - Exploits the structure of tree-based models to compute SHAP values efficiently.
  - Reduces the complexity from exponential to polynomial time by traversing the tree structure and calculating exact Shapley values in a fraction of the time required for general SHAP.

Example

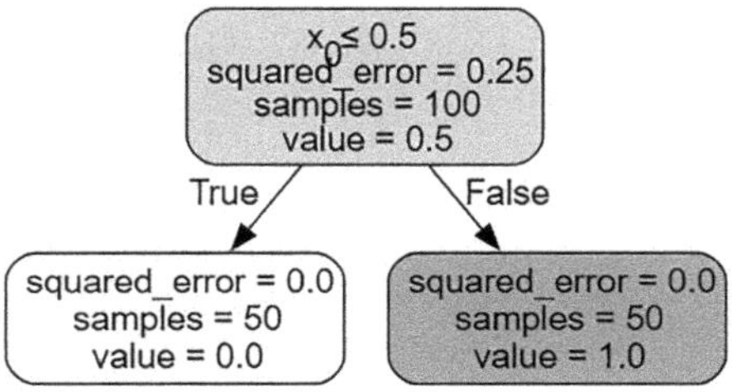

FIGURE 2.5 SHAP explanation from the above decision tree.

2.2.4 PDP Plots

PDPs provide visualizations of the relationship between a feature and the model's predicted outcome. The plots provide how average marginal effect a feature affects the model's output. PDPs show the marginal effect of one (or two) features on the predicted outcome of a machine learning model. Sample plots are given in Figures 2.6 and 2.7.

Suppose you have a prediction function $f(x)$ with features $x = (x_1, x_2, \ldots, x_p)$. If we are interested in feature(s) $S \subseteq \{1, \ldots, p\}$

- Split features into:
  - S: features of interest
  - C: all other features (complement set)

2.2.4.1 Steps to Generate PDP

For single feature based PDP visualization

- Choose a feature (e.g., Age).
- Define a grid of values (say, Age = 20, 30, 40, …).
- For each grid value:
- Replace Age with that value for all observations.
- Compute predictions.
- Average predictions.
- Plot Age vs averaged predictions.

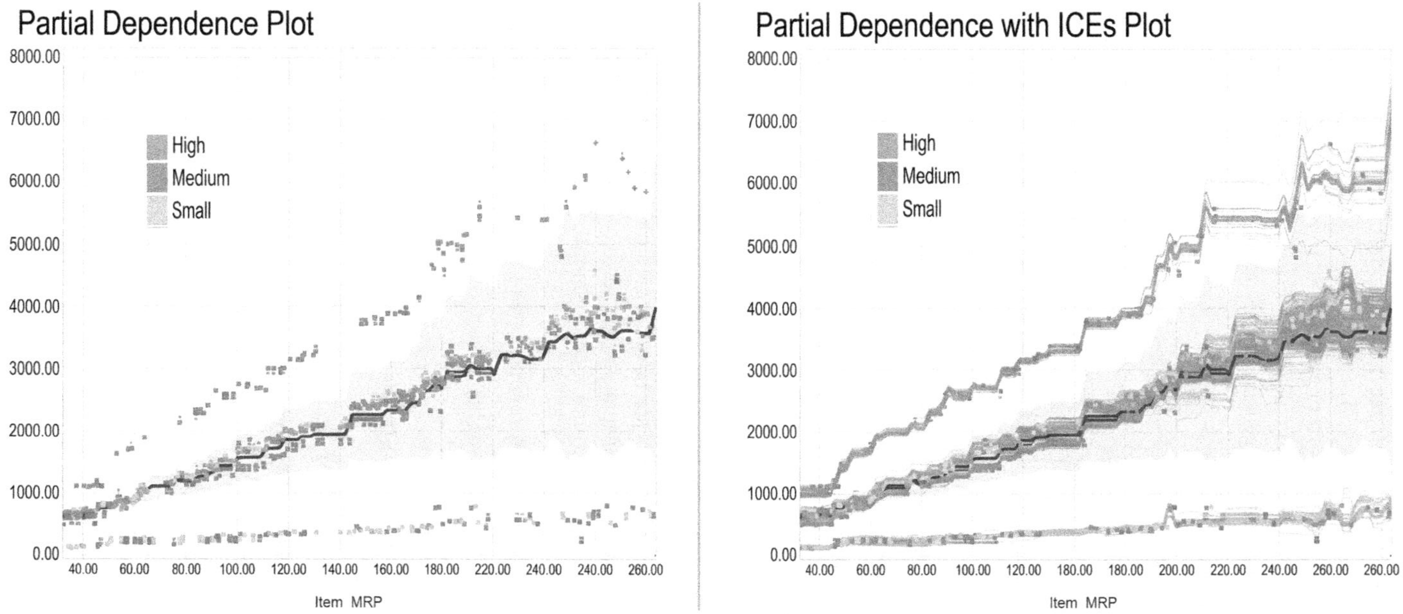

FIGURE 2.6 Partial dependence plot and PDP with ICEs plot based visualization.

(Source: https://www.knime.com/blog/apply-xai-partial-dependence-ICE)

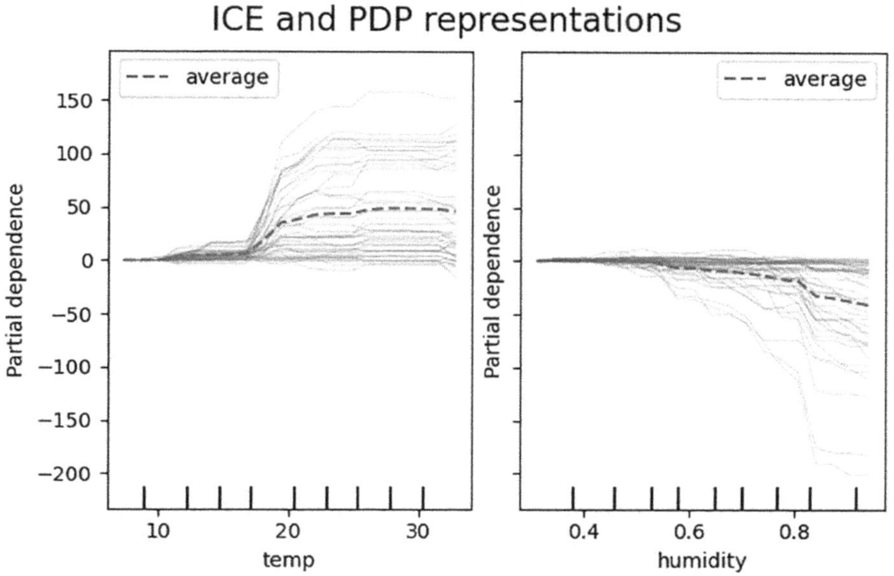

FIGURE 2.7 Partial dependence plot and PDP with ICEs plot based visualization.

(Source: https://scikit-learn.org/stable/auto_examples/inspection/plot_partial_dependence.html)

Two features based 2D surface (heatmap/contour)

- Slope: tells you how sensitive predictions are to that feature.
- Flat PDP: model is not using that feature.
- Monotonic PDP: model respects monotonic trends (e.g., higher income → higher approval probability).
- Nonlinear PDP: shows thresholds, saturation effects, or nonmonotonic behavior.

Example (loan approval with gradient boosted trees)

PDP for Income: upward slope → higher income increases approval probability.
PDP for Age: maybe flat → model does not consider age strongly.

2D PDP for Income Vs. Loan Amount:
High income & low loan → high approval
Low income & high loan → low approval

Advantages of PDPs

- It provides visualizations for any black-box model.
- The plot shows the overall feature influence on the other features, thus providing a global view.
- It is easy for regulators and business experts to get instant insight on the model's performance.
- It confirms monotonicity checks.

Drawbacks of PDPs

- PDPs generally average over marginal distribution. Hence, if a feature of interest is correlated with others, the PDP will show some unrealistic scenarios.
  A typical example is: PDP of age might evaluate Age = 10 with income = \$500K, which is unrealistic.
- PDPs won't show heterogeneity. It shows only an average effect. The subgroups will give different results.
- PDPs are expensive for large models as they need to produce high-resolution grids.
- PDPs won't give explanations on individual predictions.

2.2.5 ICE Plots

As PDPs are not providing individual explanations, the next extension is ICE which is expanded as Individual Conditional Expectation plots. They unpack the average effect into individual curves in the visualization graph.

ICE plots show the effect of a feature for each instance individually, not as an average effect as in PDP. They show how the feature's effect differs across various instances. They are particularly useful for finding interactions and heterogeneous effects. Generally, ICE plots are used together with PDP plots to show the complete explanation.

It can be represented as PDP = Mean Curve Across all ICE Curves. It is mathematically represented as,

$$\text{ICE}\_i(x\_j) = f\left(x\_j, x\_\{-j\}\wedge\{(i)\}\right)$$

where

$\text{ICE}\_i(x\_j)$ represents the predicted output of the model for the i-th instance as the value of feature

f is the trained machine learning model

Xj is the specific feature whose effect on the prediction is being investigated. Its value is systematically varied across its range or a chosen subset of its range

$x - j(i)$ represents all other features for the i-th instance, held constant at their original observed values for that specific instance.

2.2.5.1 Steps to Construct ICE Plots

1. Select a feature of interest.
2. Select the grid values of that feature.
3. For each instance in the dataset do the following:
   a. Replace the feature with each grid value with other features fixed.
   b. Predict the outcome, and the curve for that instance has to be generated.
4. Combine all the curves and prepare it as a complete plot.

Different Types of ICE Curves

- Flat ICE Curves – used to showcase some instances without considering specific features.
- Steep ICE Curves – to showcase instances that are influenced by the feature.
- Diverging / Crossing ICE Curves – They showcase the strong feature interactions and/or show the subgroup differences.
- Parallel ICE Curves – It helps to view the feature that has homogeneous effect.

Sample

Consider a loan approval system that has an ML model employed. The feature to be analyzed is taken as "income."

For this problem, the PDP visualization shows a smooth increase in approval probability with income.

With respect to ICE it shows the following visualizations:

- The curve is flat for applicants who have low income with high loan amount.
- The curve shows a sharp increase for applicants with low income and with low loan amount.
- This visualization helps to find the effect of "income" with "loan amount" in an instance-by-instance manner.

Advantages of ICE Plots

- Using ICE plots, we can get heterogeneous effects which are not usually visible in PDP plots.
- ICE plots can show the interactions between features.
- It is highly useful for subgroup fairness analysis generally taken from different populations that may respond differently.
- It is easy to implement as it is model-agnostic.

Limitations of ICE Plots

- The plots assume feature variation as realistic as PDP.
- If dataset is large, the ICE plot can get cluttered due to hundreds of curves in the plot.
- The global averaging trend taken from ICE curves that diverge strongly, the results will be misleading.

2.2.6 Rule-Based Explanations

In rule based explanations, the output is formal or symbolic rules that represent how the decisions have been taken by the model. These explanations are generated for both interpretable rule based models such as decision trees as well as from less interpretable models such as ensemble models, deep neural networks which have a long history in the decision-making process.

One such method is suggested by Cropper et al. (2021) to apply inductive logic programming that summarizes methods to automatically construct rule sets to solve a given task by taking the formal knowledge base and other guided examples.

In addition, there is an argumentation framework proposed by Vassiliades et al. (2021) to provide an interpretable logical argumentation deduces the logical background knowledge in statement form. Also, rule extraction methods aim to generate symbolic rule sets from a trained model.

Example:
Consider the same Loan Approval system which can predict whether a person is eligible for loan.

Constraints are imposed by the bank in terms of a set of rules and based on that the decision will be derived.

```
If age<60 and citizenship == true
   If income >=500000 and loan_default=='No' then
      loan_approval=true
   If income >=500000 and loan_default =='Yes' then
      loan_approval=false
If age>=60 and citizenship==false then loan_approval = false
If age<60 and citizenship==false then loan_approval=false
If age>=60 and citizenship== true
    If income >=500000 and loan_default=='No' and property==
      true then loan_approval=true
   If income >=500000 and loan_default =='Yes' and property==
      true then loan_approval=false
   If income >=500000 and loan_default=='No' and property ==
      false then loan_approval=true
   If income >=500000 and loan_default =='Yes' and property ==
      false then loan_approval=false
```

2.2.7 Counterfactual Explanations

Counterfactual explanations describe the impact of the smallest change to the feature values on the prediction outcome. They are used to give explanations of individual instances. Generally, these counterfactual explanations describe causal relationships of a scenario:

For example, "If Y had not occurred, Z would not have occurred."

Example:
Consider the same "Loan Approval System" in which the outcome of loan approval is based on the values of the input features such as age, income, citizenship, and loan_default status. To find out the important feature that primarily decides the outcome, this technique will introduce small changes to each of the feature values and find their impact on the outcome.

How it works?

To get the counterfactual explanation, the model developer will define a relevant change in the prediction of an instance of the input dataset. The counterfactual instance produces the predefined prediction as closely as possible.

Generally, it is applicable for classification type of problems in which changing the feature values will flip the class. However, if the dataset has more instances having a frequent class and only fewer instances belonging to a rare class, the small change introduced in the dataset will never flip the frequent class to the rare class.

In these cases, it is essential to find the minimal changes in the features so that the prediction probability changes from lower probability to some higher probability. Also, it is essential that the counterfactual should be as close as the original instance. To achieve it, the distance measures to find similarity such as Manhattan or Gower distance measures can be used. Another most important criterion about counterfactual explanations is that the changes being done on the features should involve fewer features. Otherwise, it will be impossible to find which feature has caused the change in the outcome as one feature's change may nullify the effect of another feature's change.

It can be done by finding the norm between the counterfactual instance and the normal instance to find the number of changed features. We can apply multiple diverse counterfactual explanations to generate different outcomes for each one.

In the Loan Approval System example, multiple counterfactual instances can be created on the features such as age, income, and loan_default status to find which one has higher impact on the outcome classification.

The explanations derived from counterfactual explanation techniques are straightforward. There are no additional or hidden changes in the procedure.

Advantages of Counterfactual Explanations

- It is very simple to implement and does not produce any harmful effects on the model's outcome.
- The advantage of counterfactual technique is it does not require to access the data or model. It needs access to the prediction function alone which can also be done through the user interface. Hence, it can be used as a trustworthy way of providing the explanations to the users of the system without showing the internal working of the model or the data.
- It offers a balance between the model predictions and the model's protection.
- It works across all Machine Learning models.

Limitations

- It is time-consuming to generate the changed instances to find the impact on outcome.
- If many features are taken for change, the effect of one feature change will nullify the effect of the other features.
- It is not applicable for imbalanced datasets having 90% of data under one class and 10% of data under another class.

2.3 MODEL-SPECIFIC

These techniques are usually associated with specific types of ML and DL models. They make use of the internal structure of the model, parameters, and learning process used by the model to generate the explanations.

For instance, the model-specific explanation techniques for different models are given below:

- Decision Trees – Rule Extraction and Decision Path visualization, which explains how splits lead to predictions.
- Linear Regression model – Coefficient interpretation and Odds Ratio, which explain the feature contribution and positive or negative impact using sign.
- Neural Network – Specific XAI methods applicable are: GradCAM, LRP, Integrated Gradients, etc. They provide visualizations in terms of feature attribution and saliency maps.
- Ensemble Models – Feature importance using impurity reduction or gain are applied as specific XAI techniques to give explanation on which features are used in ensemble decision-making.

2.3.1 LRP

LRP is an explanation technique which propagates the prediction backward using purposely designed local propagation rules. LRP is subject to the conservation property. It is designed especially for Neural Networks and Deep Learning Models. It explains the model's individual prediction by decomposing the output back to the input features by showing how much each feature contributed to the result.

For instance, we applied NN for Tumor Classification problem. It provides a result that a particular image has tumor with probability of 0.9.

Now, to examine the feature that impacted this decision, LRP redistributes this 0.9 probability score backward through the network layers until it reaches the input pixels. Each pixel then receives the relevance score Ri, that tells how strongly that pixel supported or opposed the decision or final prediction.

2.3.1.1 Core Working Principle of LRP

Each neuron passes its relevance score backward proportionally to its contribution to the next layer's activations.

There are two major steps involved:

1. Forward Pass – Input x value is fed to the trained NN model and gets an output $f(x)$ which for the sample problem gave the probability value of presence of tumor = 0.9
2. Backward Pass – Start from the output neuron's relevance score $R = f(x)$ and move backward.

For each neuron j in layer l+1, relevance is redistributed to neurons i in the previous layer l using this rule:

$$R_i = \sum j \frac{aiwij}{\sum iaiwij + \varepsilon} Rj$$

where

ai is the activation of neuron i
wij is the weight connecting neuron i to j
Rj is the relevance of neuron j
Σ is the small constant value to avoid divide by 0.

Now, the checking condition is on whether the relevance score at layer i is equal to the layer j.

$$\Sigma i\ Ri = \Sigma j\ Rj$$

2.3.2 GradCAM

GradCAM is abbreviated from Gradient-Weighted Class Activation Mapping. It is a visual explanation technique that shows the regions of an image that influence a deep learning model's outcome.

It works by combining the gradients of the target class flowing into the convolutional layer with the feature maps from that same layer.

It is model-specific as it uses CNN convolutional feature maps and gradients, and so it can't be applied to other models. It is applicable to neural network-based Deep Learning models.

The core working principle of GradCAM has the following steps:

Forward Pass – Input the image I to the trained CNN model. It outputs the predictions with output probabilities for the different classes. For instance, it gives the probability for Tumor class with 0.88 as probability value.
Backward Pass – Compute the gradient of the output probability score for the tumor class with respect to the feature maps A^k at the convolutional layer. y^{tumor}

It is given in the following equation:

$$\frac{\delta y^{\text{tumor}}}{\delta y^{\text{Ak}}}$$

Compute Feature Map Weights – For each feature map k, compute the average gradient over spatial dimensions.

Compute a weighted sum of feature maps – By applying ReLU, we can ensure that only the features positively influencing the selection of "Tumor" are highlighted.

Visualization of Results – The result of GradCAM is a 2D heatmap. We can resize it to the input image size and overlay it on top of the original image.

2.3.3 GradCAM++

GradCAM++ is an extension of the original GradCAM technique, which visualizes the regions of an input image that are important for a particular class prediction. It achieves this by utilizing the gradients of the output class with respect to the final convolutional layer of the CNN, which effectively highlights the salient areas in the image that contributed to the decision. In the implementation of GradCAM++, we calculate the gradients and apply a weighted average to obtain class activation maps. These maps are then overlaid onto the original images, providing a clear visual cue of which parts of the image the model focused on during classification. GradCAM++ offers improved localization compared to its predecessor, allowing for finer granularity in highlighting important features.

The combination of GradCAM++ with LIME and SHAP provides a comprehensive toolkit for interpreting the predictions of our third CNN model. Together, these techniques not only validate the model's reliability but also enhance our understanding of the underlying mechanisms that drive its performance.

2.4 SUMMARY

The explanations derived from XAI techniques and models are usually visual interpretations based on feature importance scores derived from simpler surrogate models or from focused data point values. XAI techniques can be categorized into four ways. The model-specific techniques are usually associated with specific types of ML and DL models. They make use of the internal structure of the model, parameters, and learning process used by the model to generate the explanations.

BIBLIOGRAPHY

Bahalul Haque, A. K. M., Islam, A. K. M. Najmul, and Mikalef, Patrick. (2023). Explainable artificial intelligence (XAI) from a user perspective: A synthesis of prior literature and problematizing avenues for future research. *Technological Forecasting and Social Change*, 186:122120. ISSN 0040-1625. https://doi.org/10.1016/j.techfore.2022.122120.

Cecilia, Panigutti, Alan, Perotti, and Dino, Pedreschi. (2020). Doctor XAI: An ontology-based approach to black-box sequential data classification explanations. In *Proceedings of the Conference on Fairness, Accountability, and Transparency (FAT'20)*, pp. 629–639.

Cecilia, Panigutti, Andrea, Beretta, Daniele, Fadda, Fosca, Giannotti, Dino, Pedreschi, Alan, Perotti, and Salvatore, Rinzivillo. (2022). Co-design of human-centered, Explainable AI for clinical decision support. *ACM Transactions on Interactive Intelligent Systems*, 13(4):1–35.

Clement, T., Kemmerzell, N., Abdelaal, M., and Amberg, M. (2023). XAIR: A systematic metareview of Explainable AI (XAI) aligned to the software development process. *Machine Learning and Knowledge Extraction*, 5(1):78–108. https://doi.org/10.3390/make5010006

Cropper, A., De Raedt, L., Evans, R., and Schmid, U. (2021). Approaches and applications of inductive programming (Dagstuhl Seminar 21192). *Dagstuhl Reports*, 11(4), 20–33.

Dino, Pedreschi, Fosca, Giannotti, Riccardo, Guidotti, Anna, Monreale, Luca, Pappalardo, Salvatore, Ruggieri, and Franco, Turini. (2018). Open the black box data-driven explanation of black box decision systems. Arxive preprint: 1806.09936.

Došilović, F. K., Brčić, M., and Hlupić, N. (2018). Explainable artificial intelligence: A survey. In *Proceedings of the 41st International Convention on Information and Communication Technology, Electronics and Microelectronics (MIPRO)*, Opatija, Croatia, pp. 210–215.

Farrow, Robert. (2023). The possibilities and limits of XAI in education: A socio-technical perspective. *Learning, Media and Technology*, 48(2):266–279.

Frąckiewicz, Marcin. (2023). Explainable AI (XAI) in education: Enhancing learning and personalized instruction. *Artificial Intelligence, TS2 Space.*

Khosravi, Hassan, Shum, Simon Buckingham, Chen, Guanliang, Conati, Cristina, Tsai, Yi-Shan, Kay, Judy, Knight, Simon, Martinez-Maldonado, Roberto, Sadiq, Shazia, and Gašević, Dragan. (2022). Explainable artificial intelligence in education. Computers and Education: Artificial Intelligence, 3:100074. ISSN 2666-920X. https://doi.org/10.1016/j.caeai.2022.100074.

Mattia, Setzu, Riccardo, Guidotti, Anna, Monreale, Franco, Turini, Dino, Pedreschi, and Fosca, Giannotti. (2021). GLocalX – From local to global explanations of black box AI models. *Artificial Intelligence*, 294:103457.

Nogueira, Ana Rita, Pugnana, Andrea, Ruggieri, Salvatore, Pedreschi, Dino, and Gama, João. (2022). Methods and tools for causal discovery and causal inference. *Wires Data Mining and Knowledge Discovery*, 12(2):e1449.

Nwakanma, C. T., Ahakonye, L. A. C., Njoku, J. N., Odirichukwu, J. C., Okolie, S. A., Uzondu, C., Ndubuisi Nweke, C. C., and Kim, D.-S. (2023). Explainable artificial intelligence (XAI) for intrusion detection and mitigation in intelligent connected vehicles: A review. *Applied Sciences*, 13(3):1252. https://doi.org/10.3390/app13031252

Orestis, Lampridis, Riccardo, Guidotti, and Salvatore, Ruggieri. (2021). Explaining sentiment classification with synthetic exemplars and counter-exemplars, discovery science. In *International Conference on Discovery Science*, pp. 357–373. Springer, Cham.

Riccardo, Guidotti, Anna, Monreale, Fosca, Giannotti, Dino, Pedreschi, Salvatore, Ruggieri, and Franco, Turini. (2021). Factual and counterfactual explanations for black box decision making. *IEEE Intelligent Systems*, 34(6):14–23.

Riccardo, Guidotti, Anna, Monreale, Salvatore, Ruggieri, Franco, Turini, Fosca, Giannotti, and Dino, Pedreschi. (2022). A survey of methods for explaining black box models. *ACM Computing Surveys (CSUR)*, 51(5):1–42.

Saeed, Waddah, and Omlin, Christian. (2023). Explainable AI (XAI): A systematic meta-survey of current challenges and future opportunities. *Knowledge-Based Systems*, 263:110273. ISSN 0950-7051. https://doi.org/10.1016/j.knosys.2023.110273

Theissler, Andreas, Spinnato, Francesco, Schlegel, Udo, and Guidotti, Riccardo. (2022). Explainable AI for time series classification: A review, taxonomy and research directions. *IEEE Access*, 10:100700–100724.

Tjoa, Erico, and Guan, Cuntai. (2020). A survey on explainable artificial intelligence (XAI): Towards medical XAI. *IEEE Transactions on Neural Networks and Learning Systems*, 32(11):4793–4813.

Vassiliades, A., Bassiliades, N., and Patkos, T. (2021). Argumentation and explainable artificial intelligence: A survey. *The Knowledge Engineering Review*, 36: e5.

3 Intrinsically Interpretable Models

3.1 INTRODUCTION

The ML models that have interpretability in their design itself are classified as intrinsically interpretable models. It is generally defined by the level of ML algorithm applied. In these models, the search is constrained so that they are interpretable in nature.

For instance, the Mean Squared Error usually used to analyze the fitness of regression models is self-explanatory and the algorithm will provide insight on whether the model fits linearly according to the features.

Hence, the models which are self-interpretable because of their algorithmic design are called intrinsically interpretable or inherently interpretable models.

Some of the intrinsically interpretable models are:

- Linear Regression
- Logistic Regression
- Decision Trees
- Rule-based Models
- ProtoViT
- Ensemble Boost Trees
- GAMs

Based on their scope of interpretability, these intrinsically interpretable models may be fully interpretable, partially interpretable, and individually interpretable. Hence, by analyzing their scope of interpretability, the above models will fit under any one of these categories.

Characteristics of Intrinsically Interpretable Models

- Easy to debug
- Complete insight on internal working
- Well-defined justification of outputs on inputs
- Human in the loop for easy verification of the model
- Consistent with domain knowledge
- Bridges the gap between accuracy and interpretability

Let us now see the elaborated explanations on each of the intrinsically interpretable models listed above.

DOI: 10.1201/9781003630241-4

3.2 REGRESSION MODEL INTERPRETABILITY

Linear regression and Logistic regression models are white-box models in nature. In these models, it is easy to identify the cause-and-effect relationships. The least squares method used in these models is well understood and they have algorithmic-level transparency. The linear regression model performs the regression task whereas the logistic regression model performs the classification task. The mathematical model of these models is given below:

$$y = w_0 + w_1 x_1 + \cdots w_n x_n$$
$$= w_0 \sum_{i=1}^{n} w_i x_i \tag{3.1}$$

For binary classification problems, Logistic regression model is applied. It generally models the prediction factors and the probability of outcome using the logit function such as sigmoid function. This function converts the calculated values to probabilities. Based on that, the class label is provided for the prediction result.

$$f(x) = 1/\left(1 + e^\wedge(-x)\right) \tag{3.2}$$

Consider Boston Housing dataset which has 14 features and has 506 records.

To find the house price from features that have real and continuous values, these regression models help in finding the target output.

It is diagrammatically given in Figure 3.1.

3.2.1 Heat Map

Correlation of input with output is given by means of a heat map, which provides the visualization on multicollinearity. This provides insight on which feature is highly correlated with one or more features in the given dataset. It is given in Figure 3.2.

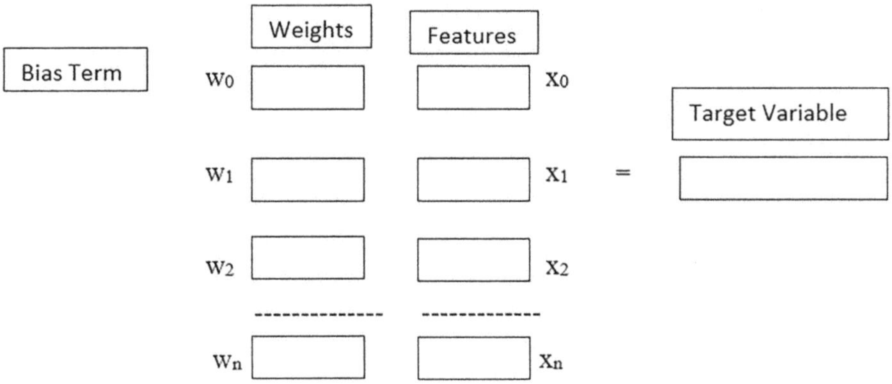

FIGURE 3.1 Regression model representation.

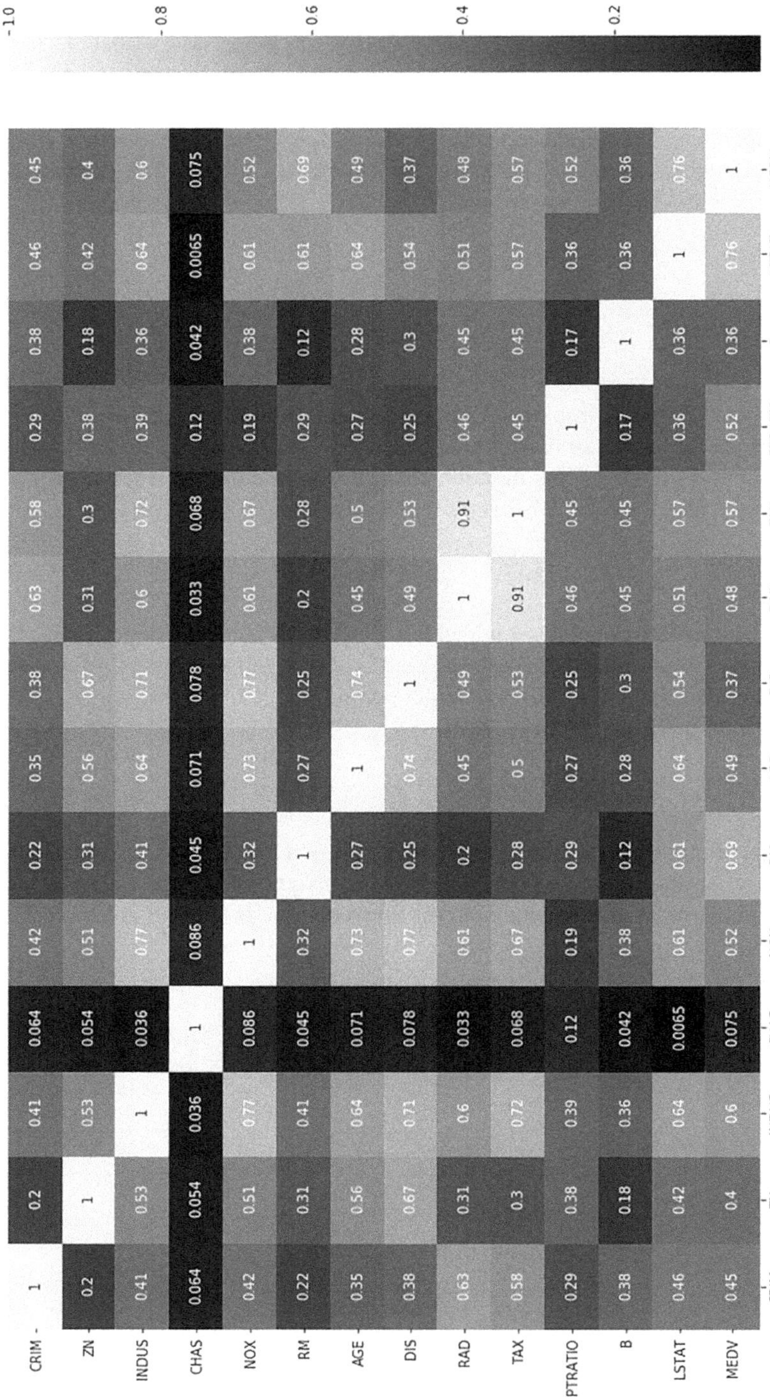

FIGURE 3.2 Heatmap visualization of features correlation.

3.2.2 Interpretation from Heatmap

From the heatmap visualization given in Figure 3.2, it is understood that the correlation coefficient value ranges between -1 to +1. If the correlation value is nearer to +1, it indicates that there is a positive correlation between two variables. If it is nearer to -1, it indicates a strong negative correlation between the two variables.

From the above visualization, it is understood that feature RM has a higher positive correlation with the outcome variable price with a value of 0.7.

But the feature LSTAT has a higher negative correlation with the outcome variable price with a value of -0.74.

The features such as RAD and TAX have correlation values of 0.91. This indicates these pairs are strongly correlated with one another. Similarly, the pair of features DIS and AGE have a negative correlation value of -0.75.

This indicates that when we are selecting features for a regression model, it is essential to check for multicollinearity.

3.2.3 Code Snippet of Linear Regression Model

```
#Train Test split
X,y=load_dataset('BostonHousing.csv')
X_train,X_test,y_train,y_test=train_test_split(X,y,test_
  size=0.33,random_state=1)
print('Train',X_train.shape,y_train.shape)
print('Test',X_test.shape,y_test.shape)

Train (339, 13) (339,)
Test (167, 13) (167,)
from sklearn.linear_model import LinearRegression
#model building
model=LinearRegression()
model.fit(X_train,y_train)
```

```
▾ LinearRegression
LinearRegression()
```

```
model score
rsq=model.score(X_train,y_train)
print("R-square value",rsq)
R-square value 0.7217346524579973
#model parameters
print('w0',model.intercept_)
print('w1',model.coef_)
w0 39.68280792749965
w1 [-9.91995332e-026.27806786e-027.25812126e-023.01077411e+00
 -2.06617880e+013.44496595e+003.78004138e-03 -1.44827470e+00
```

```
3.09305140e-01 -1.16485640e-02 -9.46002716e-017.20017685e-03
 -5.28368116e-01]

#Evaluate the model
MSE=mean_squared_error(y_test,ypred)
RMSE=math.sqrt(MSE)
print('Mean squared Error',MSE)
print('Root Mean squared Error',RMSE)
Mean squared Error 20.698475744484366
Root Mean squared Error 4.549557752626552
MAE=mean_absolute_error(y_test,ypred)
print('Mean Absolute Error',MAE)
Mean Absolute Error 3.4174722788016596
```

The number of features present in a dataset represents the dimensions of the input space. Hence, this white-box interpretable model has a weighted sum of the input features. By plotting the graph between target and input, the importance of a feature can be analyzed. The MAP value provides the relative difference for the error expressed in percentage (%) error. More insights can be derived from positive weight, negative weight, and their impact on linear monotonic output.

3.2.4 Other Visualizations Derived from Regression Models

3.2.4.1 Feature Importance Plot

From the feature importance plot, one can derive the features that have largest negative impact and multicollinearity which shows how an input feature is highly correlated with one or more other features. A sample derived from the following code is given in Figure 3.3.

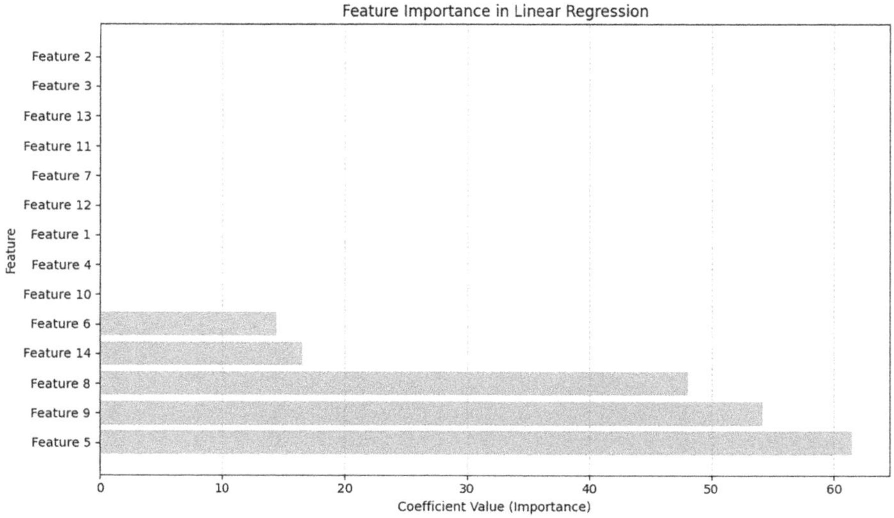

FIGURE 3.3 Sample feature importance plot.

```
from sklearn.datasets import make_regression
from sklearn.linear_model import LinearRegression
import matplotlib.pyplot as plt
import pandas as pd
import numpy as np

# 1. Generate a sample dataset (or load your own data)
X, y = make_regression(n_samples=1000, n_features=14, n_
   informative=5, random_state=42)
feature_names = [f"Feature {i+1}" for i in range(X.shape[1])]
# Create feature names for plotting
# Convert to DataFrame for easier handling of feature names
X_df = pd.DataFrame(X, columns=feature_names)

# 2. Train a Linear Regression model
model = LinearRegression()
model.fit(X_df, y)

# 3. Get feature importance (coefficients)
# In linear regression, the absolute value of the coefficients
   directly indicates feature importance.
importance = model.coef_

# 4. Create a DataFrame for plotting
feature_importance_df = pd.DataFrame({'Feature': feature_
   names, 'Importance': importance})

# Sort by absolute importance for better visualization
feature_importance_df['Absolute_Importance'] =
   np.abs(feature_importance_df['Importance'])
feature_importance_df = feature_importance_df.sort_
   values(by='Absolute_Importance', ascending=False)

# 5. Plot feature importance
plt.figure(figsize=(10, 6))
plt.barh(feature_importance_df['Feature'], feature_importance_
   df['Importance'], color='skyblue')
plt.xlabel('Coefficient Value (Importance)')
plt.ylabel('Feature')
plt.title('Feature Importance in Linear Regression')
plt.grid(axis='x', linestyle='--', alpha=0.7)
plt.tight_layout()
plt.show()
```

3.2.4.2 Correlation Plots

The correlation plots help to identify whether the results are lying along the straight line of the regression model. Using these plots, one can easily get insights on the various issues present in the given dataset. The sample correlation plots are given in Figure 3.4.

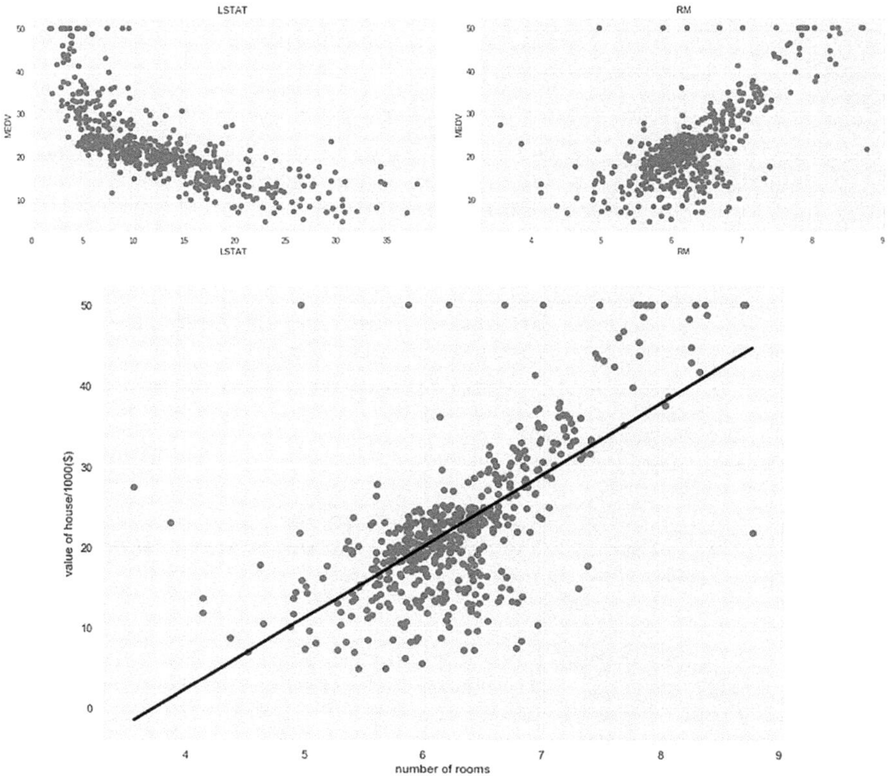

FIGURE 3.4 Correlation plots for the sample dataset.

Limitations of Linear Regression

- If the relationship between input and output is nonlinear, it will be difficult to interpret the outcome.
- There is no way to find whether the model's prediction result is underfitting and the dataset has bias.

3.3 DECISION TREES BASED INTERPRETABILITY

Decision trees can be used to model complex nonlinear relationships. They can be used for both nonlinear monotonic and nonlinear nonmonotonic. Here, nonmonotonic means a change in the input could result in a change in the output in any direction and at a varying rate. And monotonic means a change in the input results in a change in the output in the same direction.

Decision Tree model is used for both classification and regression. The methodology of decision tree is to recursively split the given data based on "Information Gain," a value calculated from "Gini Index" or "Entropy."

By applying a recursive cut, the tree-like structure will be created and is used to make a decision. The internal nodes in a decision tree contribute a decision and each outcome is taken from the leaf node in that branch.

$$\text{Gini index} = 1 - \Sigma\left(p_i\right)^\wedge 2 \tag{3.3}$$

$$\text{Entropy}\left(s\right) = -\Sigma\left(p_i \log_2\left(p_i\right)\right) \tag{3.4}$$

Decision trees can be applied for both regression based on MSE and classification based on Gini impurity and Entropy. It is a highly interpretable model.

A sample decision tree is shown below (see Figure 3.5):

```
Corresponding explanations are provided as part of the code
   execution.
[Text(1729.8, 978.48, 'insulin <= 127.5\nentropy = 0.927\
   nsamples = 514\nvalue = [338, 176]\nclass = 0'),
   Text(669.5999999999999, 761.0400000000001,
'skinthick <= 22.5\nentropy = 0.722\nsamples = 325\nvalue =
   [260, 65]\nclass = 0'), Text(446.4, 543.6, 'entropy = 0.0\
   nsamples = 67\nvalue = [67, 0]\nclass = 0'), Text(892.8,
   543.6,
'bp <= 26.45\nentropy = 0.814\nsamples = 258\nvalue = [193,
   65]\nclass = 0'), Text(446.4, 326.1600000000001, 'bp <=
   9.1\nentropy = 0.28\nsamples = 62\nvalue = [59, 3]\nclass
   = 0'), Text(223.2, 108.72000000000003, '\n (…) \n'),
   Text(669.5999999999999, 108.72000000000003, '\n (…) \n'),
   Text(1339.1999999999998, 326.1600000000001,
'pedigree <= 0.634\nentropy = 0.9\nsamples = 196\nvalue =
   [134, 62]\nclass = 0'), Text(1116.0, 108.72000000000003,
   '\n (…) \n'), Text(1562.3999999999999, 108.72000000000003,
   '\n (…) \n'), Text(2790.0, 761.0400000000001, 'bp <= 29.85\
   nentropy = 0.978\nsamples = 189\nvalue = [78, 111]\nclass =
   1'), Text(2008.8, 543.6, 'pregnant <= 1.5\nentropy = 0.874\
   nsamples = 51\nvalue = [36, 15]\nclass = 0'), Text(1785.6,
   326.1600000000001, 'entropy = 0.0\nsamples = 13\nvalue =
   [13, 0]\nclass = 0'), Text(2232.0, 326.1600000000001,
'bmi <= 73.0\nentropy = 0.968\nsamples = 38\nvalue = [23,
   15]\nclass = 0'), Text(2008.8, 108.72000000000003, '\n
   (…) \n'), Text(2455.2, 108.72000000000003, '\n (…) \n'),
   Text(3571.2, 543.6,
'insulin <= 158.5\nentropy = 0.887\nsamples = 138\nvalue
   = [42, 96]\nclass = 1'), Text(3124.7999999999997,
   326.1600000000001,
'bmi <= 61.0\nentropy = 0.98\nsamples = 84\nvalue = [35,
   49]\nclass = 1'), Text(2901.6, 108.72000000000003, '\n
   (…) \n'), Text(3348.0, 108.72000000000003, '\n (…) \n'),
   Text(4017.6, 326.1600000000001,
'skinthick <= 30.5\nentropy = 0.556\nsamples = 54\nvalue
   = [7, 47]\nclass = 1'), Text(3794.3999999999996,
   108.72000000000003, '\n (…) \n'), Text(4240.8,
   108.72000000000003, '\n (…) \n')]
```

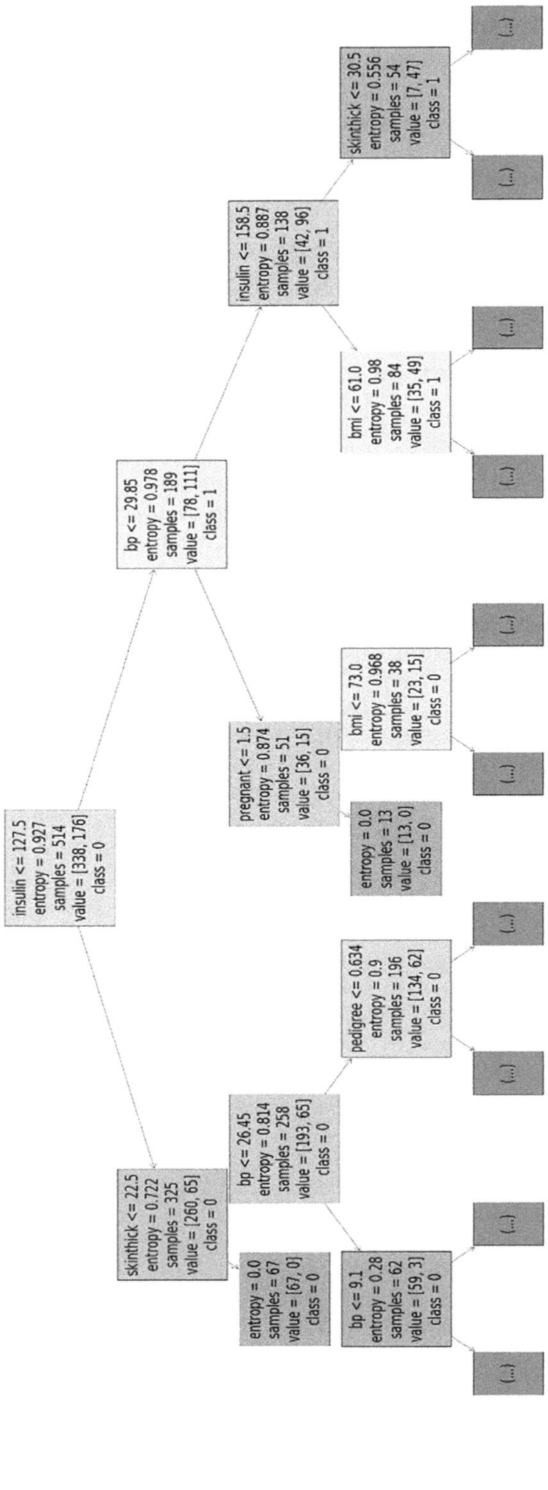

FIGURE 3.5 Decision tree visualization for each node and branching information.

Feature importance visualization from a decision tree can be derived using the following steps:

Step 1. Compute the importance of a node in the binary tree.

- The importance of a node is computed as in equation (3.5)

$$I_k^n = p_k \cdot m_k - p_k^l \cdot M_k^l - p_k^r \cdot M_k^r \tag{3.5}$$

where

I_k^n is importance of node k
p_k is the proportion of samples to reach node k
m_k is the impurity measure of node k
p_k^l is the proportion of samples to reach left subtree of node k
M_k^l is the impurity measure of left subtree of node k
p_k^r is the proportion of samples to reach right subtree of node k
M_k^r is the impurity measure of right subtree of node k

The decrease in the cost function or impurity measure for that node weighted by the probability of reaching that node in the tree.

Step 2. Summing up the importance of the nodes that split on that feature normalized by the importance of all the nodes to 100 as given in equation (3.6).

$$I_i^f = \frac{\sum_{j\varepsilon J} I_j}{\sum_{k\varepsilon K} I_k} \tag{3.6}$$

where

I_i^f is the importance of feature f
$\sum_{j\varepsilon J} I_j$ is the sum of importance of all nodes j that split at feature i
$\sum_{j\varepsilon J} I_j$ is the sum of importance of all nodes k in the decision tree

3.3.1 Feature Importance Plot from Decision Tree

From the above plot given in Figure 3.6, prune the feature with lowest importance to improve performance of model.

Limitations of Decision Trees

- Problem of overfitting and the model is said to have high variance.
- The change in input will ask to change the model and affects output.
- Hyper-parameter tuning – Max_depth, Minimum Sample for node.

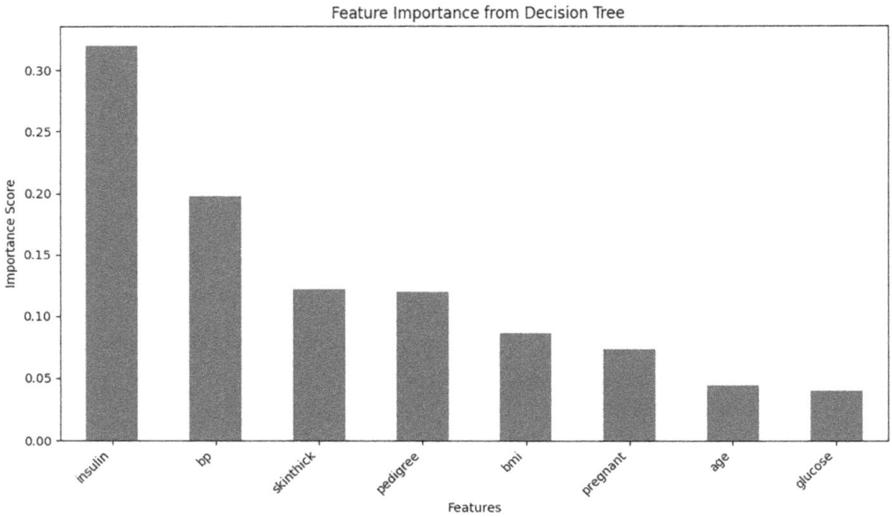

FIGURE 3.6 Feature importance plot to improve the performance of the model.

3.4 GAMS

Generalized Additive Model (GAM) is a nonlinear model. Here, the nonlinear relationships are modeled with smoothing functions for each of the features and then it will add all of them. Each feature has its own associated smoothing function that best models the relationship between that feature and the target. It is represented as,

$$y = w_0 + f_1(x_1) + f_2(x_2) + \ldots + f_n(x_n) \tag{3.7}$$

where

$f_1(x_1)$ is the smoothing function for feature x_1
$f_2(x_2)$ is the smoothing function for feature x_2
$f_n(x_n)$ is the smoothing function for feature x_n.

Here, regression splines are widely used as smoothing functions. They are practical and computationally efficient. It provides a weighted sum of linear or polynomial functions. These polynomial functions are also known as basis functions. A sample smoothing function is given below:

$$\mathrm{fj}^{(\mathrm{x_j})} = \sum_{k=1}^{K} w_k b_k(x_j) \tag{3.8}$$

where

$\sum_{k=1}^{K} w_k b_k(x_j)$ is the smoothing function represented as a weighted sum of basis functions.

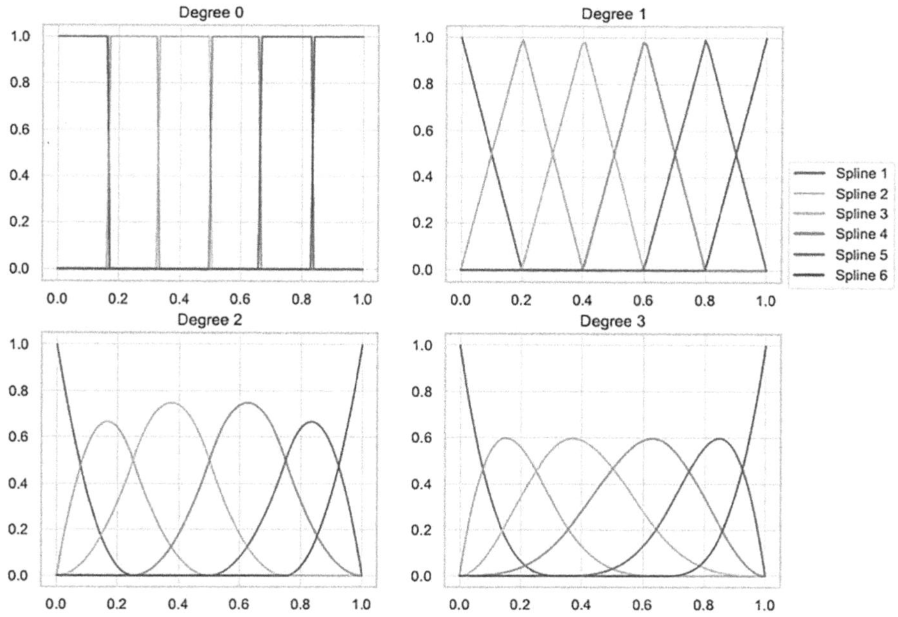

FIGURE 3.7 Sample splines using GAM.

In this function, w_k represents the weight of function and b_k represents the base function.

3.4.1 Interpretability Through GAM Visualizations

Here, a spline is used as a polynomial of degree with continuous derivatives. It is shown in Figure 3.7.

In GAM, a regression spline is a weighted sum of basis functions or splines. This is essentially a linear regression problem to learn the weights. It uses the method of least squares and gradient descent. It is needed to specify hyper-parameter which represents the number of knots or degrees of freedom. Here, K splines will have $K+3$ knots/degrees of freedom. It determines hyper-parameter using cross-validation.

3.4.2 Sample Code for GAM

```
from pygam import LinearGAM, s, f
from pygam.datasets import wage
X, y = wage(return_X_y=True)
# GAM model
gam = LinearGAM(s(0) + s(1) + f(2))
gam.gridsearch(X, y)
# plotting
plt.figure();
fig, axs = plt.subplots(1,3);
```

```
titles = ['year', 'age', 'education']
for i, ax in enumerate(axs):
    XX = gam.generate_X_grid(term=i)
    ax.plot(XX[:, i], gam.partial_dependence(term=i, X=XX))
    ax.plot(XX[:, i], gam.partial_dependence(term=i, X=XX,
       width=.95)[1], c='r', ls='--')
    if i == 0:
        ax.set_ylim(-30,30)
        ax.set_title(titles[i]);
```

The GradCAM visualization is given in Figure 3.8.

```
from pygam import LinearGAM
from pygam.datasets import mcycle
X, y = mcycle(return_X_y=True)
gam = LinearGAM(n_splines=25).gridsearch(X, y)
XX = gam.generate_X_grid(term=0, n=500)
plt.plot(XX, gam.predict(XX), 'r--')
plt.plot(XX, gam.prediction_intervals(XX, width=.95),
   color='b', ls='--')
plt.scatter(X, y, facecolor='gray', edgecolors='none')
plt.title('95% prediction interval');
```

The GAM generated prediction intervals plot is given in Figure 3.9.

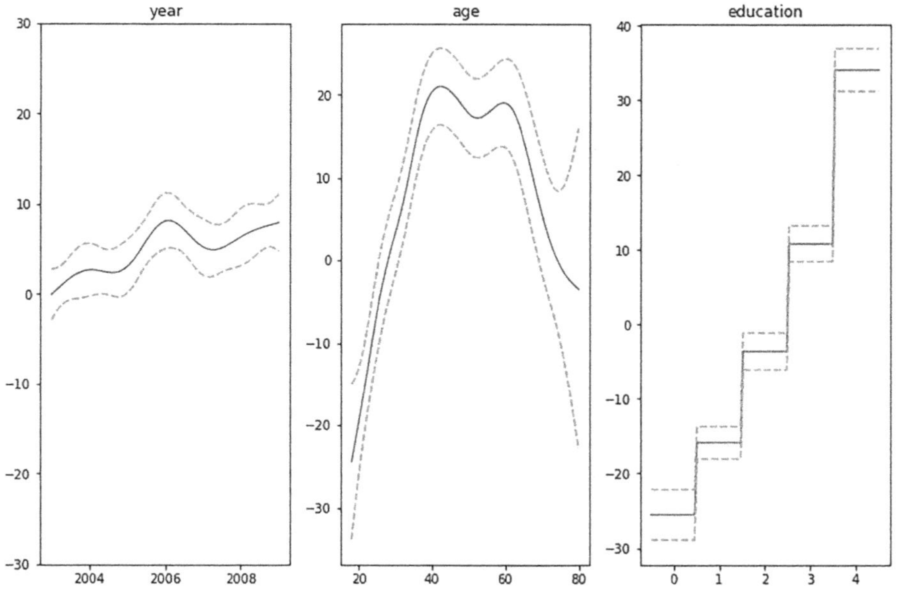

FIGURE 3.8 GAM model based visualization.

(Source: https://pygam.readthedocs.io/en/latest/notebooks/tour_of_pygam.html)

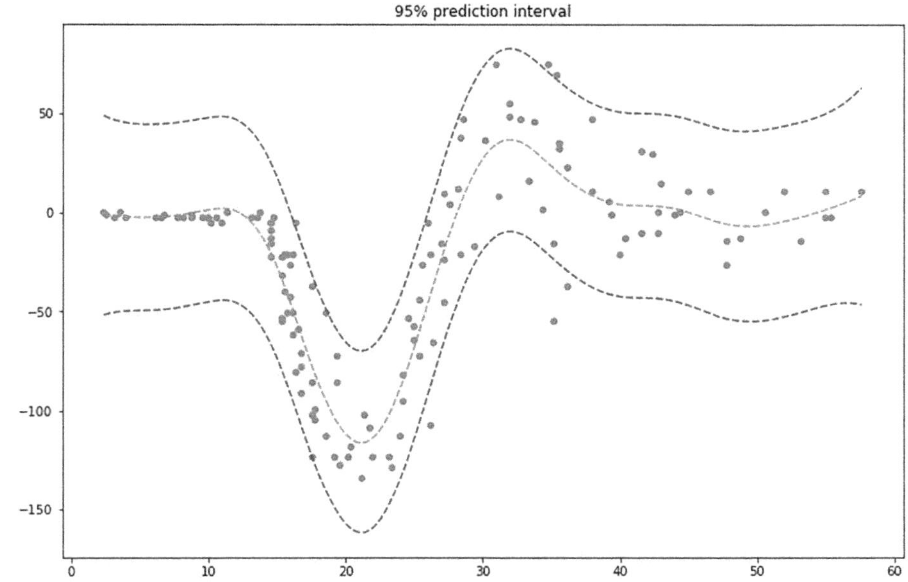

FIGURE 3.9 Prediction intervals plot using GAM.

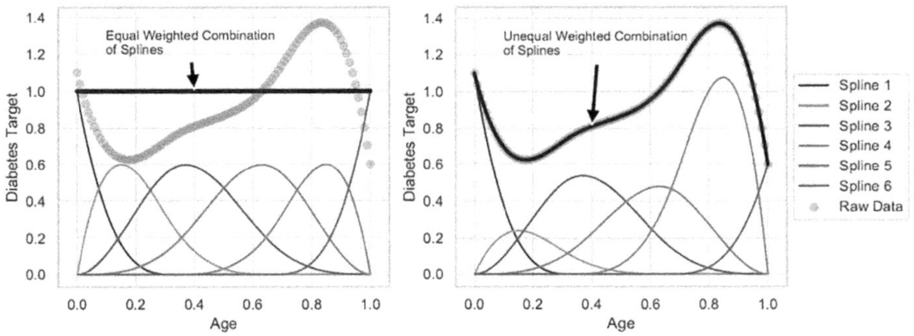

FIGURE 3.10 A sample degree-3 or cubic splines.

Degree-3 or cubic splines are used in several applications, as they can capture most general shapes. In order to capture a general shape, we will take a weighted sum of the splines.

A sample cubic spline visualization of degree-3 is given in Figure 3.10.

Equal weight for 6 splines gives – 1 in first graph.

3.4.3 Interpreting the Results of GAM

It provides the visualization of the effects of each feature on the target. The partial dependence or partial effects plots for the interpretation of GAM results are widely applied. The PDP plots (partial dependence plots) look at the effect of each feature by marginalizing on the rest.

It is highly interpretable as it provides the average effect of each feature value on the target variable. We can see whether the target response to the feature is linear or nonlinear or monotonic or nonmonotonic.

Limitations of GAM

- There may be overfitting occurring in GAMs, but can be resolved with regularization.
- GAM is sensitive to feature values outside the range in the training set.
- GAM may have reduced predictive power when exposed to outlier values.
- GAMs have limited predictive power for critical tasks.

3.5 RULE-BASED MODELS

A rule-based model represents the decision logic as a collection of if-then rules in human-readable format. These models exhibit the working of expert reasoning and help to explain why a certain prediction is taken. They are interpretable, auditable, trustworthy, and can provide both local as well as global predictions.

Each rule has a specific pattern on the data which leads to a particular outcome.

IF condition_1 AND condition_2 THEN class=1

The rule-based models provide transparent reasoning steps.

Example:

Let us consider a healthcare prediction model to find whether a person is at high risk of cardiac arrest. Then rules can be framed as below:

Rule 1: IF age>45 AND BMI>30 THEN risk = HIGH
Rule 2: IF bp>120 THEN risk = HIGH
Rule 3: IF age<35 AND Exercise=regular THEN risk=LOW
Rule 4: IF ECG=irregular AND bp<120 THEN risk=LOW

The interpretation is very clear based on the rules on why a patient is classified as high risk or low risk.

Advantages

- Easily interpretable
- Reasoning chain is transparent.
- Bias or fairness validation-based audit is possible
- Particularly useful for problem domains with constraints-based decision-making.

Limitations

- Rules make complex relationships oversimplified which leads to erroneous decision-making.
- For high-dimensional data, the scalability of the rule base is very difficult.
- Conflicts and redundancies of rules appear.
- Not highly accurate due to oversimplification.

3.6 PROTOVIT

ProtoViT stands for Prototype-based Vision Transformer. It is an interpretable vision transformer model which combines the power of transformers with prototype learning to make predictions more transparent and understandable.

Generally, it is applied for image processing applications. It explains why an image belongs to a particular class by comparing image patches to learned prototypes that represent the visual patterns of each class.

Let

- Zi = embedding of image patch i
- pj = prototype j (learned vector)

Then similarity between patch i and prototype j is computed as in equation (3.9).

$$s_{ij} = e^{(-\|zi-pj\|^2)} \tag{3.9}$$

The final class score is calculated as in equation (3.10).

$$y_c = \sum_{j\Sigma c} \max_i (s_{ij}) \tag{3.10}$$

where

c is a class and each class score is the **sum of max similarities** between image patches and class prototypes – this can be traced back and visualized directly.

3.6.1 Working Methodology

The ProtoViT model introduces interpretable prototypes into the transformer architecture. Each prototype represents a typical visual concept learned from the training data. During inference, the model does the following steps:

1. It divides the input image into patch embeddings.
2. It compares each patch embedding with stored prototypes.
3. It then identifies the prototype which closely resembles the image.
4. Finally, it gives the classification decision based on the prototype similarity.

3.6.2 Sample ProtoViT Model

Consider ProtoViT model applied for healthcare analytics.

The given problem is to classify whether a person has cancer or not using MRI scan images by classifying the images into Glioma, Meningioma, or Pituitary Cancer. The following phases are done for the prediction:

Step 1. Training Phase – ProtoViT learns both feature representations via transformer layers and prototypes used as representative embeddings for comparing and predicting the class.

Prototype P1 → typical texture of Glioma (irregular bright edges)
Prototype P2 → smooth pattern of Meningioma
Prototype P3 → central round lesion (Pituitary)

Step 2. Inference Phase – When a new MRI scan image is given as input, this phase will do the following functions:

Split the image into 16 × 16 patches which are embedded by ViT encoder.
Compare each patch embedding to all stored prototypes.
Compute similarity score.
Aggregate similarity scores to find the appropriate class.

Patch#5 → similar to Prototype P1 (0.89)
Patch#8 → similar to Prototype P1 (0.84)
Patch#12 → similar to Prototype P2 (0.32)

Step 3. Explainability Output Phase – The model highlights the patches that matched P1 and classified it as Glioma. Similarly, the other two prototypes will be used to classify the input image into the appropriate class.

Prediction: Glioma (0.92 confidence)
Matched Prototypes: P1 (0.89), P3 (0.05)
Explanation: The tumor's bright irregular boundary resembles prototype P1 of Glioma.

Advantages

1. It gives both classification and prediction.
2. It provides the prototype-based explanation along with predicted class.

Limitations

1. Complete dependency on prototypes for decision-making – It will be purely dependent on the quality and diversity of the prototypes.
2. Because of the integration of prototype layer in the ViT model, the computational complexity will be higher.
3. Limited scalability for large datasets.
4. There is difficulty in balancing semantic interpretability and discriminative power.

3.7 ENSEMBLE BOOST TREES

Ensemble Boost Trees (EBT) refer to models that combine many small decision trees into a strong predictive model using boosting techniques. They are ensembles of decision trees trained sequentially in which each tree focuses on correcting the errors produced by the previous models.

It is represented as in equation (3.11).

$$\hat{y} = \sum_{m=1}^{M} \alpha_m h_m(x) \tag{3.11}$$

where

$h_m(x)$ is the prediction from tree m
α_m is the weight assigned to that tree.
The final prediction = weighted sum of predictions from all trees.

It provides interpretations through each decision tree which is a part of the EBT. Using this model, one can get interpretations on which features are important in decision-making along with how boosting iterations refine the decision logic.

3.7.1 Working Methodology

The EBT model aims to achieve both performance and explainability. The steps are given below:

1. Start with a weak learner which is a single decision tree.
2. Measure its errors.
3. Train a new tree to fix the errors.
4. Repeat the above procedure till performance is not improving.

3.7.2 Sample EBT Model

Medical diagnosis to predict whether a person has diabetes or not.

The features in the dataset are: Glucose_level, BMI, Age, BP, Insulin.

Step 1. Train an Ensemble Boost Tree (e.g., XGBoost).
It builds 100 small trees sequentially, each improving upon the previous one.

Step 2. Prediction for a Sample
Suppose we test a new patient:

Age = 50
Glucose_Level = 180
BMI = 32
BloodPressure = 85
Insulin = 140

The model predicts:

Diabetes_Risk = 0.87 (High)

Using Tree Explainer which is normally represented using SHAP, the interpretations are derived and it is shown in Table 3.1.

TABLE 3.1
SHAP Values for Each Feature and Interpretation

Feature	SHAP Value	Impact on Risk
Glucose_Level (180)	+0.32	Strong positive effect
BMI (32)	+0.21	Moderate positive effect
Age (50)	+0.15	Small positive effect
BloodPressure (85)	+0.01	Minimal impact
Insulin (140)	−0.02	Slightly reduces risk

3.7.2.1 Global Feature Importance Plot (from XGBoost or SHAP)

From the application of SHAP on XGBoost model on the given dataset, the importance score of each feature is calculated. It is given in Table 3.2.

TABLE 3.2
Global Feature Importance Values

Feature	Importance
Glucose_Level	0.45
BMI	0.30
Age	0.15
Insulin	0.07
BloodPressure	0.03

Advantages

1. High predictive accuracy
2. Interpretability at feature level
3. Interpretations are derived directly from model internals
4. Provides both local and global insights
5. It can handle mixed data

Limitations

1. Model complexity
2. At times, feature importance can be misleading if features are correlated.
3. It needs some explainability tools such as SHAP to give the interpretations.
4. It normally provides approximate explanations only as it uses SHAPley values-based interpretation.

3.8 INTRINSICALLY INTERPRETABLE ARCHITECTURES

3.8.1 ProtoPNet

ProtoPNet (Prototype-based Part Network) is a deep neural network architecture that introduces human-understandable prototype reasoning into CNN-based image classification. Instead of classifying an image purely through black-box features, ProtoPNet asks:

"Does this image contain parts that look like prototypical parts of each class?"

$$\text{ProtoPNet} = \text{CNN} + \text{Prototype Layer} + \text{Linear Classifier}$$

It explains predictions through visual prototypes that represent conceptual parts (like "wing of a bird," "ear of a dog," or "bright region of tumor").

3.8.1.1 Working Methodology

Traditional CNNs learn hidden features that are not human-interpretable.

ProtoPNet adds a prototype layer that stores visual representations (patches) of each class learned during training.

During inference:

- The model extracts feature maps from the input image (via CNN).
- It compares every patch of the image with the learned prototypes.
- The prediction is based on how similar parts of the input image are to the class-specific prototypes.

This gives both high accuracy and visual explanations: which parts of the image resemble which learned prototypes.

The sequence of operations performed in this architecture is given in Figure 3.11:

3.8.1.2 Interpretability Mechanism

ProtoPNet's interpretability comes from prototype-based reasoning.

Reasoning Style:

"This image is a cat because this patch of fur looks like the prototype fur patch of cats."

So the explanation is example-based, visual, and semantically meaningful.

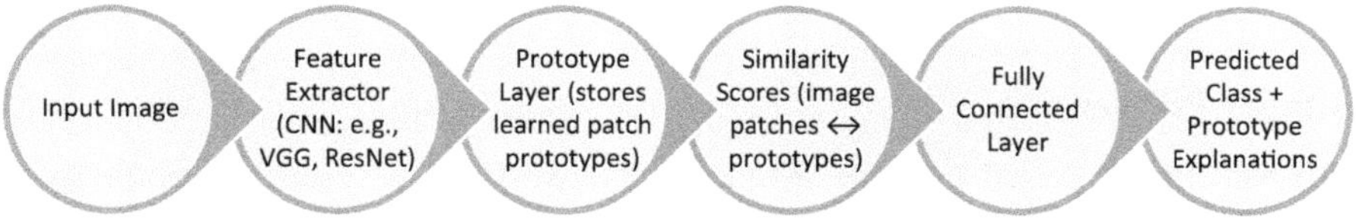

FIGURE 3.11 ProtoPNet architecture.

Each prediction can be explained as:

- Which prototypes activated strongly.
- Which regions of the image triggered those activations.

Visualized side by side with prototype examples.

Let
$f(x)$ = feature map of image x
pj = prototype j
$d(f(x), pj)$ is the distance between image patch and prototype.

Each prototype pj belongs to a specific class c_j.

Then the class score is calculated using equation (3.12).

$$\sum_{j \cdot \varepsilon c} w_{c_j} \cdot \max^{\left(-\|zi - pj\|^2\right)} \tag{3.12}$$

This means each class gets a score based on the most similar patch in the image to each of its prototypes.

During inference, ProtoPNet produces.

- Activation Heatmap which highlights image regions that matched prototypes.
- Prototype Patch Visualization which shows corresponding prototype examples.
- Similarity Score which quantifies strength of each match.

Advantages
1. Human-interpretable evidence
2. Prototype interpretation is embedded in the model's decision pipeline itself
3. Visual transparency to find which region leads to the prediction.
4. Good accuracy and interpretability

Limitations
1. Predictions and interpretations are purely based on the prototypes.
2. Computational overhead is very high.
3. Limited scalability of prototype size.

3.8.2 PIPNet

PIP-Net ("Patch-based Intuitive Prototypes Network") is a recent prototype (part-prototype)-based model designed to improve both classification and interpretability in image tasks. PIP-Net was introduced by Nauta et al. in CVPR (2023) as an

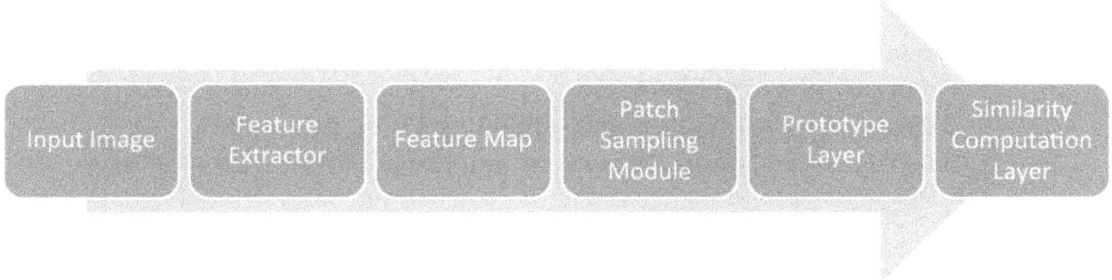

FIGURE 3.12 PiPNet architecture.

interpretable deep learning model based on part prototypes. It extends prototype networks (like ProtoPNet) by learning intuitive, patch-level, nonredundant, and human-interpretable prototypes.

The sequence of operations is given in Figure 3.12:

3.8.2.1 Working Methodology

Step 1. Training Phase

1.1. Feature learning
The backbone CNN/ViT is trained to extract discriminative patch features. Standard cross-entropy loss for classification.

1.2. Prototype learning
Each prototype is optimized to at least one patch from its class (minimize distance to its best-matching patch) and has to be distant from patches of other classes (encourages class-specificity). The objective function ensures each prototype represents a unique, intuitive visual part of its class.

1.3. Regularization for interpretability
To ensure human interpretability, several losses are added as shown in Table 3.3.

Step 2. Inference Phase
For a new image:

- Extract patch embeddings via backbone.
- Compare each patch to all prototypes.

TABLE 3.3
Losses Added in PiPNet

Loss Term	Purpose
Diversity Loss	Prevents multiple prototypes from learning the same pattern.
Compactness Loss	Ensures prototype stays close to real training patches (not abstract noise).
Sparsity Loss	Forces few prototypes to contribute per class → concise explanations.
Orthogonality Loss	Encourages prototypes to capture distinct parts.

- Identify top-matching patches for each prototype.
- Weighted similarity scores are combined to predict class.
- Generate visual explanation:
- Highlight image regions corresponding to top prototypes.
- Display corresponding prototype patches and their class associations.

Where, sim(): cosine similarity or negative Euclidean distance, w_{cj}: sparse class-prototype weights, and s_j: prototype activation strength.

This formulation makes interpretability intrinsic – the model's internal scoring directly corresponds to visible prototype activations.

3.9 EVALUATION OF INTRINSICALLY INTERPRETABLE MODELS

3.9.1 Predictive Models' Performance Metrics

Even though the model is interpretable, it must still perform competitively. The various metrics and their purposes are given in Table 3.4.

3.9.2 Faithfulness Evaluation

The metrics which are associated with measuring how truly the explanations reflect the model's internal reasoning are given in Table 3.5.

TABLE 3.4
Performance Metrics and Their Purposes

Metric	Purpose	Example
Accuracy/F1-score	General performance on test data	Classification tasks
AUC/ROC	Discrimination ability	Medical diagnosis
Precision–Recall	Useful for imbalanced datasets	Anomaly or disease detection
Confidence Calibration	Whether prediction probabilities reflect true likelihoods	Expected Calibration Error (ECE)

TABLE 3.5
Faithfulness Measures

Metric	Measurement	Method Applied
Fidelity/Faithfulness	Agreement between model output and explanation model	Compare prototype activations vs. decision importance
Completeness	Whether explanations capture all relevant factors	Fraction of output variance explained by interpretable parts
Monotonicity Test	Whether removing influential interpretable elements changes predictions as expected	Removing top prototypes should lower confidence
Causal Consistency	Explanations should remain consistent under small perturbations	Perturbation-based sensitivity tests

3.10 INTRINSIC INTERPRETABILITY FOR ETHICAL AI

Ethical AI refers to the development and deployment of AI systems that are having ethical principles as given in Table 3.6.

Intrinsic interpretability directly supports transparency, accountability, fairness, and trustworthiness, which are the core pillars of ethical AI. It is listed in Table 3.7.

The mechanism applied by the intrinsically interpretable models and their ethical benefits are listed in Table 3.8.

TABLE 3.6
Ethical Principles in XAI

Ethical Principle	Description
Transparency	The system's functioning and reasoning are open and understandable.
Accountability	Decisions can be traced to specific causes or responsible entities.
Fairness & Nondiscrimination	Model treats individuals or groups without bias.
Reliability & Safety	Predictions are robust, consistent, and do not cause harm.
Privacy & Autonomy	Protects user data and respects human decision-making.
Trustworthiness	Humans can rely on the system's explanations and behavior.

TABLE 3.7
Ethical AI Principles vs. Intrinsic Interpretability

Ethical AI Principle	Intrinsic Interpretability	Evidence
Transparency	The reasoning steps are visible and understandable by humans. No "black box."	Nature (Caruana et al., 2015): Interpretable GAMs used in pneumonia risk prediction prevented unsafe clinical recommendations.
Accountability	Decisions can be traced to specific prototypes, rules, or features.	ProtoPNet (Chen et al., 2019, CVPR): Each decision is linked to visual prototypes ("looks like this region"), allowing clear justification.
Fairness & Bias Detection	Enables inspection of which features or prototypes influence predictions, making biases discoverable.	Morales-Forero et al., 2025: Found that interpretable prototypes expose racial bias in dermatology models where spurious features were used.
Trust & Human-AI Collaboration	Explanations improve trust and human oversight, especially in high-risk domains like healthcare or law.	PIP-Net (Nauta et al., 2023, CVPR): Medical experts rated explanations 4.2/5 for alignment with clinical reasoning, improving adoption trust.
Safety & Reliability	Faithful explanations help detect model failures early and prevent unsafe actions.	Rudin, 2019 (Nature Machine Intelligence): Argues black-box models should not be used for criminal justice or healthcare due to unverifiable safety.
Legal & Regulatory Compliance	Intrinsic interpretability helps satisfy "right to explanation" clauses in data protection laws.	EU AI Act (2024) and GDPR Article 22: Mandate explainability for automated decision systems; interpretable models are inherently compliant.

TABLE 3.8
Interpretability Mechanism vs. Ethical Benefit

Mechanism	Description	Ethical Benefit
Visibility of internal reasoning	Exposes model's decision factors.	Enables transparency and fairness auditing.
Bias localization	Reveals biased features (e.g., race, gender).	Supports fairness interventions.
Causal reasoning	Shows how cause–effect relationships drive decisions.	Improves accountability and scientific validity.
Right to Explanation	Provides human-understandable reasons for decisions.	Fulfills ethical and legal requirements.
User empowerment	Users can contest or question model outputs.	Promotes autonomy and justice.

TABLE 3.9
Ethical AI Pipeline

Step	Component	Ethical Benefit
Data Preprocessing	Fair sampling, data audit	Reduces input bias
Model Choice	ProtoPNet or PIP-Net	Built-in transparency
Decision Tracking	Prototype activations logged	Accountability and audit trail
Human-in-the-loop Review	Expert verification of prototypes	Safety and trust
Deployment Monitoring	Stability of explanations checked	Long-term ethical compliance

The Ethical AI Pipeline using Intrinsically Interpretable Model and the corresponding ethical benefits are listed in Table 3.9.

3.11 CHALLENGES IN INTRINSIC INTERPRETABILITY

We can categorize the challenges into five major domains as given in Table 3.10.

There is no universally accepted metric for interpretability. Unlike accuracy or F1-score, interpretability is subjective, multidimensional, and task-dependent.

3.11.1 Dimensions of Evaluation

The various dimensions used to evaluate the intrinsically interpretable models and how the evaluation procedure will check are given in Table 3.11.

TABLE 3.10
Challenges in Intrinsic Interpretability

Challenge Type	Description	Example
Trade-off Challenge	Balancing accuracy with simplicity	Decision trees become complex in high dimensions
Cognitive Challenge	Human understanding varies; interpretability is not equal to understandability	Clinicians may misread prototypes
C. Design & Scalability Challenge	Building interpretable models for large or high-dimensional data	Deep models > millions of parameters
D. Evaluation Challenge	Measuring interpretability objectively is difficult	Lack of standardized metrics
E. Ethical & Practical Challenge	Interpretability may lead to information overload or false confidence	Misinterpretation of explanations

TABLE 3.11
Dimensions of Evaluation vs. Interpretability

Dimension	Evaluation Test	Example
Transparency	Is the reasoning visible?	Tree visualization
Faithfulness	Does the explanation reflect true model logic?	Gradient correlation
Simplicity	Is the explanation concise?	Rule length
Human trust/usability	Do users understand and act on it?	User studies
Stability	Do explanations remain consistent?	Perturbation tests

3.12 SUMMARY

This chapter has dealt with several intrinsically interpretable models with their application on ethical AI. These models aim to achieve a balance between predictive performance and human interpretability, allowing users to trace how input features contribute to the final output. Some of the typical intrinsically interpretable models are regression models, decision trees, rule-based models, GAMs, etc. They are very easy to audit and validate for fairness, bias, and errors. They are useful in high-stakes domains (e.g., healthcare, finance, law). They facilitate trust and accountability in AI systems. Within this framework, intrinsically interpretable models play a crucial role because they make their decision logic transparent by design, allowing stakeholders to understand, trust, and govern AI behavior.

Intrinsically interpretable models are essential tools for building ethical and trustworthy AI systems. They provide clarity, fairness, and accountability by making AI decisions understandable, auditable, and justifiable, which aligns directly with the ethical principles of transparency, nonmaleficence, and human autonomy.

REVIEW QUESTIONS

Remember/Understand

1. Define intrinsically interpretable models.
2. Differentiate between posthoc explainability and intrinsically interpretable models.
3. List any four models under intrinsically interpretable models.
4. List advantages and disadvantages of GAM model.
5. Compare decision tree–based visualization and rule-based visualization.

Apply

1. Consider the following sample data taken from Melbourne housing dataset:

Suburb	Rooms	Type	Dist from Busstand	Bedroom2	Bathroom	Car_ parkit	Landsize	Building Area	Year Built	Price
Abbotsford	2	h	2.5	2	1	1	202			1480000
Abbotsford	2	h	2.5	2	1	0	156	79	1900	1035000
Abbotsford	3	h	2.5	3	2	0	134	150	1900	1465000
Abbotsford	3	h	2.5	3	2	1	94			850000
Abbotsford	4	h	2.5	3	1	2	120	142	2014	1600000
Abbotsford	2	h	2.5	2	1	0	181			941000
Abbotsford	3h	h	2.5	4	2	0	245	210	1910	1876000

The output (dependent) feature is price.

(a) Consider Mr. Joe who owned a house and the details of his house are in row 6 of the above dataset. He wants to increase the price of the house. Provide him support by means of counterfactual explanations to get insight on it.

(b) Identify the causal relationship between the features and the output and draw the causal relationship model as a graph using the derived causal relationships. (5 Marks)

2. Consider a medical insurance dataset that has independent features as age, sex, BMI, number of children, smoking, and region and it has to provide decision on how much to be charged as insurance cost which has some specific ranges based on stated criteria. Total records in the dataset are: 1338; training set has 66% and test set has 33%.

(a) Identify the appropriate ML algorithm to be applied for it. In order to provide interpretation to the decision provided by the model based on impurity measure, you are asked to showcase how node importance and feature importance can be calculated.
(b) The same dataset is analyzed using linear regression model and the following statistics are arrived after regression. Using R-squared and t-statistic value provide the interpretation of the model's decision.

 (a) Examine how GAM is an intrinsically interpretable ML model with the various smoothing functions used to represent nonlinear data. (6 Marks)

OLS Regression Results

Dep. Variable:	y	R-squared:	0.746
Model:	OLS	Adj. R-squared:	0.743
Method:	Least Squares	F-statistic:	325.0
Date:	Mon, 15 May 2023	Prob (F-statistic):	1.33e-257
Time:	18:17:28	Log-Likelihood:	-9081.7
No. Observations:	896	AIC:	1.818e+04
Df Residuals:	887	BIC:	1.822e+04
Df Model:	8		
Covariance Type:	nonrobust		

	coef	std err	t	P > | t |	[0.025	0.975]
age	261.5684	14.590	17.929	0.000	232.934	290.202
bmi	347.0973	35.963	9.651	0.000	276.514	417.680
children	371.7622	170.842	2.176	0.030	36.461	707.064
sex	121.1237	411.092	0.295	0.768	-685.703	927.950
smoke	2.37e+04	505.206	46.914	0.000	2.27e+04	2.47e+04
northwest	-339.6184	575.367	-0.590	0.555	-1468.857	789.621
southeast	.886.4996	592.536	-1.496	0.135	-2049.436	276.437
southwest	-803.8848	589.730	-1.363	0.173	-1961.313	353.543
intercept	-1.243e+04	1228.046	-10.119	0.000	-1.48e+04	-le+04

Analyze

(a) Examine how GAM is an intrinsically interpretable ML model with the various €smoothing functions used to represent nonlinear data.
(b) For diabetes analytics system, GAM model is applied to find the features that have high effect on diabetes. The corresponding plots are given for your reference: Provide the interpretation appropriately.

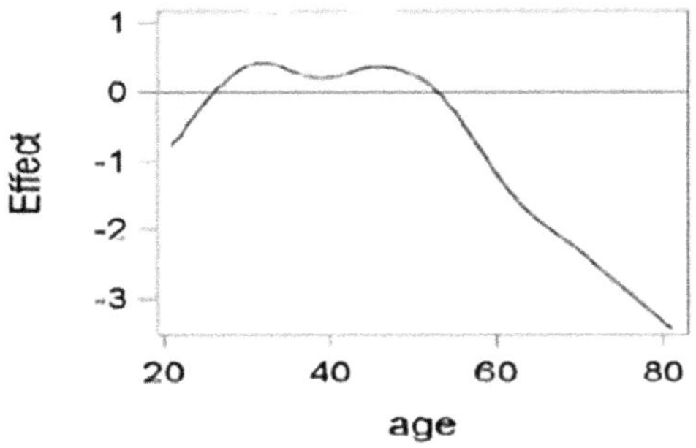

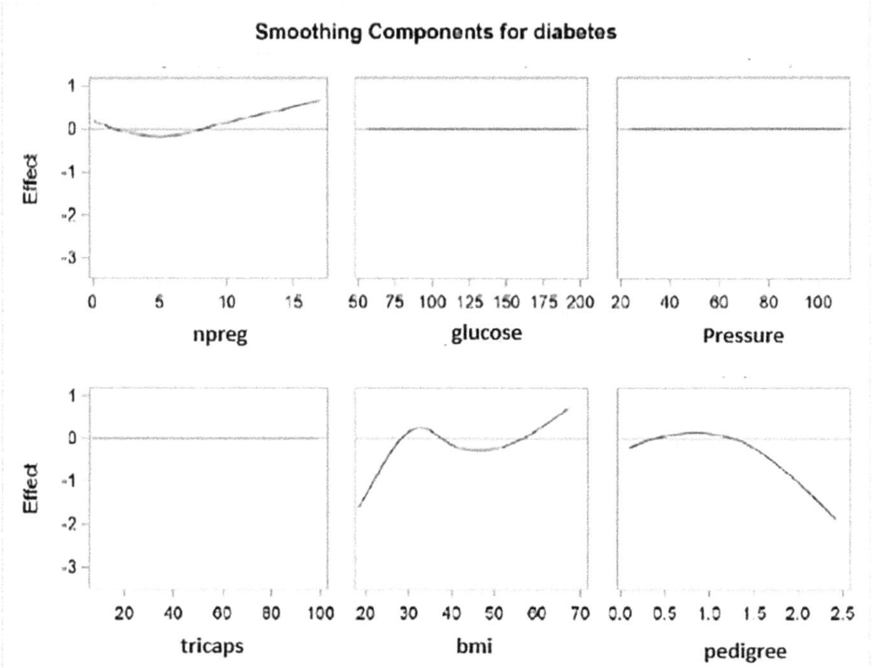

BIBLIOGRAPHY

Agarwal, R., Melnick, L., Frosst, N., Zhang, X., Lengerich, B., Caruana, R., and Hinton, G. (2021). Neural additive models: Interpretable machine learning with neural nets. *Advances in Neural Information Processing Systems (NeurIPS)*, 34:4699–4711.

Alvarez-Melis, D., and Jaakkola, T. S. (2018). Towards robust interpretability with self-explaining neural networks. arXiv:1806.07538.

Angelino, E., Larus-Stone, N., Alabi, D., Seltzer, M., and Rudin, C. (2017). Learning certifiably optimal rule lists. In *Proceedings of 23rd ACM SIGKDD International Conference on Knowledge Discovery and Data Mining (KDD)*, pp. 35–44.

Angelino, E., Larus-Stone, N., Alabi, D., Seltzer, M., and Rudin, C. (2018). Learning certifiably optimal rule lists for categorical data. *Journal of Machine Learning Research*, 18(234):1–78.

Bertsimas, D., and Dunn, J. (2017). Optimal classification trees. *Machine Learning*, 106(7):1039–1082.

Caruana, R., Lou, Y., Gehrke, J., Koch, P., Sturm, M., and Elhadad, N. (2015). Intelligible high-accuracy models in healthcare: Predicting pneumonia risk and hospital 30-day readmission. In *Proceedings of 21st ACM SIGKDD International Conference on Knowledge Discovery and Data Mining (KDD)*, Sydney, Australia, pp. 1721–1730.

Chen, C., Li, O., Tao, C., Barnett, A., Su, J., and Rudin, C. (2019). This looks like that: Deep learning for interpretable image recognition. *Advances in Neural Information Processing Systems (NeurIPS)*, 32:8928–8939.

Doshi-Velez, F., and Kim, B. (2017). Towards a rigorous science of interpretable machine learning. arXiv preprint arXiv:1702.08608.

Friedman, J. H., and Popescu, B. E. (2008). Predictive learning via rule ensembles. *The Annals of Applied Statistics*, 2(3):916–954.

Hastie, T., and Tibshirani, R. (1990). *Generalized Additive Models*. London, UK: Chapman & Hall/CRC Press.

Letham, B., Rudin, C., McCormick, T. H., and Madigan, D. (2015). Interpretable classifiers using rules and Bayesian analysis: Building a better stroke prediction model. *The Annals of Applied Statistics*, 9(3):1350–1371.

Lou, Y., Caruana, R., Gehrke, J., and Hooker, G. (2013). Accurate intelligible models with pairwise interactions. In *Proceedings of 19th ACM SIGKDD International Conference on Knowledge Discovery and Data Mining (KDD)*, Chicago, IL, pp. 623–631.

Morales-Forero, A., Rueda, L. J., Herrera, R., Bassetto, S., and Coatanea, E. (2025). Predictive representativity: Uncovering racial bias in AI-based skin cancer detection. arXiv preprint arXiv:2507.14176.

Nauta, M., Schlötterer, J., Van Keulen, M., and Seifert, C. (2023). Pip-net: Patch-based intuitive prototypes for interpretable image classification. In *Proceedings of the IEEE/CVF Conference on Computer Vision and Pattern Recognition*, pp. 2744–2753.

Ribeiro, M. K. T., Singh, S., and Guestrin, C. (2016). Why should I trust you? Explaining the predictions of any classifier. In *Proceedings of 22nd ACM SIGKDD International Conference on Knowledge Discovery and Data Mining (KDD)*, pp. 1135–1144.

Rudin, C. (2019). Stop explaining black box machine learning models for high-stakes decisions and use interpretable models instead. *Nature Machine Intelligence*, 1(5):206–215.

Rudin, C., Chen, C., Chen, Z., Huang, H., Semenova, L., and Zhong, C. (2022). Interpretable machine learning: Fundamental principles and 10 grand challenges. *Statistics Surveys*, 16:1–85.

Sudjianto, A., and Zhang, A. (2021). Designing inherently interpretable machine learning models. In *Proceedings of ACM ICAIF 2021 Workshop on Explainable AI in Finance*, arXiv:2111.01743.

Wang, T., Rudin, T., Doshi-Velez, F., Liu, Y., Klampfl, E., and MacNeille, P. (2017). A Bayesian framework for learning rule sets for interpretable classification. *Journal of Machine Learning Research*, 18(70):1–37.

Yang, H., Rudin, C., and Seltzer, M. (2016). Scalable Bayesian rule lists. arXiv:1602.08610 (also: Proceedings of the 34th AAAI/related proceedings, 2017 preprints).

4 Posthoc Explainability Models

4.1 INTRODUCTION

For real-time complex problem domains, the application of Deep Learning (DL) models is of higher-level need. When we employ these DL models, due to their black-box nature, one couldn't understand why and how a particular decision is being taken. When we generally say explanation of the output or outcome, we will simply explain what that output is all about and not about the reasons behind such outcome.

By employing XAI systems, the required information on justification of results can be derived which will be useful when we have unexpected decisions being derived. It further helps to prove that the model is trustworthy and follows ethical guidelines. By understanding how the system works, one can get the behavioral aspects of the system which provides higher visibility on vulnerabilities in the system and can be rectified on getting such insights.

For DL models, "posthoc explainability" based XAI techniques are applied to reveal the internal working of the model. In this chapter, we will see the different types of posthoc explainability approaches and how they are applied in real-time case studies.

In general, posthoc explainability methods help to provide explanations from black-box models such as ensemble models and DL models. The outcome from such methods will help to get insight on how the model has taken a specific decision.

4.2 POSTHOC EXPLAINABILITY METHODS

Posthoc explainability methods are "Model Agnostic." There are basically two types of Posthoc explainability approaches available.

- Local Explanations – They focus on granular level of the DL system. In the case of data, they provide individual explanations on each instance of data and provide trust on model's outcomes.
- Global Explanations – They focus on the overall model's outcome and provide an insight on how the model's decision-making process works.

4.3 TYPES OF POSTHOC EXPLANATION APPROACHES

According to the way of their working, the posthoc explainability approaches are classified into the following categories:

DOI: 10.1201/9781003630241-5

- Explanations based on Feature Importance
- Rule-based explanations
- Pixel Attribution based explanations
- Prototype-based explanations
- Counterfactual explanations

4.3.1 Explanations Based on Feature Importance

The XAI methods such as LIME and SHAP provide explanations based on feature importance. LIME gives the explanations on the model's decision based on its learning locally around the predictions. The explanations given by LIME are trustworthy in which it indicates both positive and negative impacts assessment of each feature on the DL model's outcome. The visualizations are given in Figures 4.1–4.3.

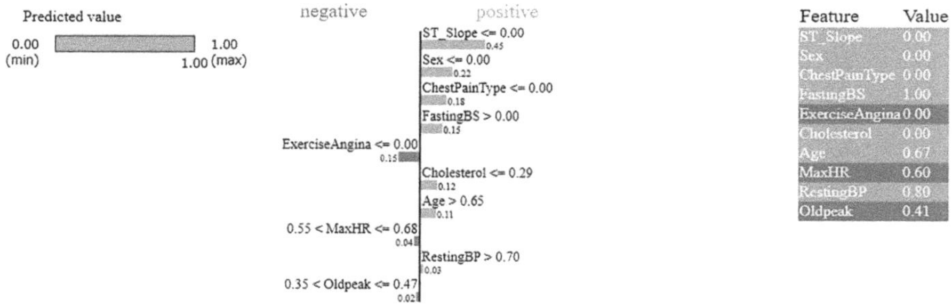

FIGURE 4.1 LIME plot sample with feature importance in positive and negative impacts.

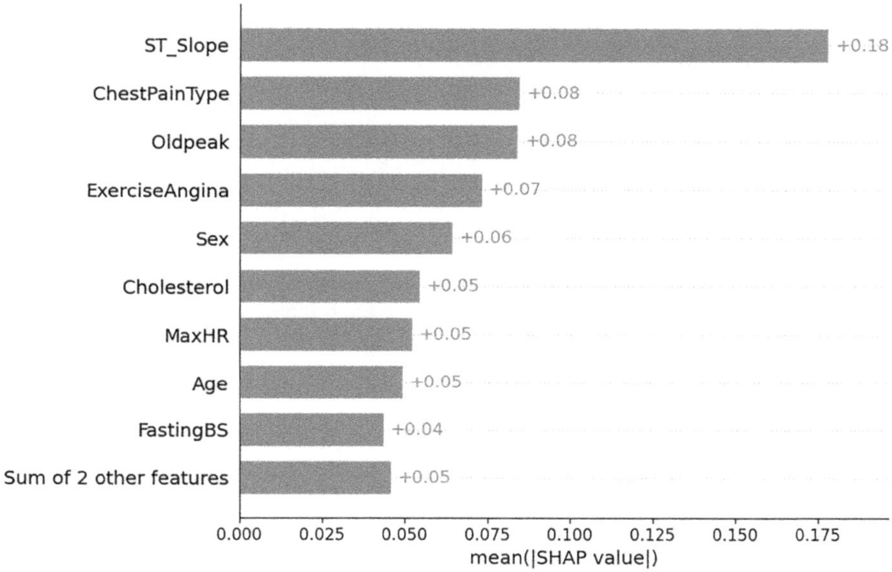

FIGURE 4.2 Sample SHAP plot.

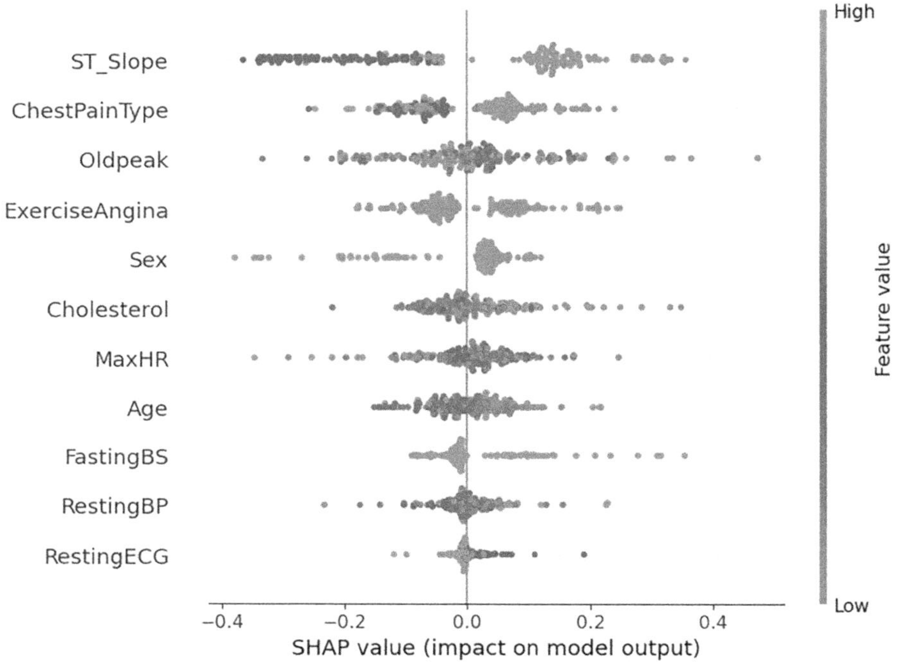

FIGURE 4.3 SHAP plot for feature impact on model output.

4.3.1.1 LIME Code

```
from xgboost import XGBClassifier

data = pd.read_csv(fn)
print("Heart Data")
data.head(10)

explain = explainer.explain_instance(data_row = x_test.
   iloc[0], predict_fn = xgb.predict)
explain.show_in_notebook(show_table = True)
```

The other XAI method that focuses on feature importance–based explanation is SHAP. This method assigns a SHAPley score to each feature based on its impact on the output for a particular prediction with reference to other features.

4.3.1.2 SHAP Code

```
pip install shap
import shap
shap_explainer = shap.Explainer(xgb.predict, x_test)
shap_values = shap_explainer(x_test)
shap.plots.bar(shap_values)
shap.summary_plot(shap_values)
```

4.3.2 Rule-Based Explanations

Rule-based explanations express the model's behavior by means of if-then rules in human-readable form.

In posthoc explainability, these rules are extracted from the trained model after the model is trained using the dataset. Some of the DL models that have this kind of explanations are: DNNs, Random Forest Classifiers, Gradient Boosting Methods, and Support Vector Machine classifier.

Steps involved in this type are listed below:

Step 1. Train a black-box model with given dataset.
Step 2. Get model predictions.
Step 3. Analyze the behavior of the model in terms of input and output.
Step 4. Extract the rules from the trained model that acts as a surrogate of the actual model.
Step 5. By generating the if-then rules, the explanations are generated from the model.

Rule-based explanations are of two types: (i) Global Rule-based explanations which provide overall model behavior and (ii) Local Rule-based explanations which provide explanations for each individual prediction.

Sample

Consider a "Home Loan Approval System." Say for instance, Gradient Boosting-based black-box model is applied for prediction. The dataset has several features such as Name, Address, Credit_Score, Income, defaulter_loan_history with outcome variable loan_sanction_status.

Now, based on the model's outcome, the rules extracted for a given test sample of Credit_Score=800; Income=700000; defaulter_loan_history=Yes, the model gave the prediction result as "No."

```
Rules are:
Rule 1:
IF Credit_Score >= 750
AND Income >=500000
AND defaulter_loan_histroy= No
THEN load_sanction_status = Yes
Rule 2:
IF Credit_Score >= 750
AND Income >=500000
AND defaulter_loan_histroy= Yes
THEN load_sanction_status = No
Rule 3:
IF Credit_Score <750
AND Income >=500000
AND defaulter_loan_histroy= Yes
THEN load_sanction_status = No
...
```

Likewise, all the rules will be extracted. Using these rules, one can easily understand how the model has taken the decision.

Some of the techniques applied in rule-based explanations are:

1. Decision Tree Surrogates – mimic complete black-box model.
2. Anchors – They provide local rules.
3. RuleFit – It is a linear model with extracted rules.
4. LIME Rules – converts LIME explanations into rules.

Advantages

- They provide high-precision outcome from the black-box models. The interpretability of the explanations generated as IF-THEN rules will be high when compared to other visualizations.
- They provide global explanations with logical reasoning.

4.4 ANCHOR EXPLANATIONS

Anchor method provides high-precision rule-based explanations in a human-understandable way. These are if-then conditions which act as "anchors" which guarantee a prediction with high precision irrespective of the values of the other features.

Sample code for anchor:

```
import numpy as np
import pandas as pd
from sklearn.datasets import fetch_openml
from sklearn.model_selection import train_test_split
from sklearn.ensemble import RandomForestClassifier
from sklearn.preprocessing import OrdinalEncoder
from alibi.explainers import AnchorTabular
#Load dataset for Anchor Explanation
data = load_breast_cancer()
X = pd.DataFrame(data.data, columns=data.feature_names)
y = data.target
X_train, X_test, y_train, y_test = train_test_split(X.values,
   y, test_size=0.2, random_state=42)
model = RandomForestClassifier(n_estimators=150,
   random_state=42)
model.fit(X_train, y_train)

feature_names = X.columns.tolist()
explainer = AnchorTabular( predictor=model.predict, feature_
   names=feature_names )
explainer.fit( X_train, disc_perc=(25, 50, 75))
instance = X_test[5]
# Get model prediction
prediction = model.predict(instance.reshape(1, -1))[0]
```

```
# Generate anchor explanation
explanation = explainer.explain(instance)
print("Model Prediction:", "Benign" if prediction == 1 else
   "Malignant")
print("\nAnchor Rules:")
for rule in explanation.anchor:
    print(" -", rule)
print("\nPrecision:", explanation.precision)
print("Coverage:", explanation.coverage)
```

Output:

```
Model Prediction: Malignant
Anchor Rules:
 worst radius > 16.2
 worst concave points > 0.17
 mean texture > 20.5
Precision: 0.97
Coverage: 0.18
```

4.4.1 Pixel-Attribution-Based Explanations

These XAI methods provide explanations on a black-box model's prediction by assigning an importance score to each pixel of an image. Using this, one can get information on which pixels in the image have influenced the model's outcome. These methods are especially applicable for DL models focusing on image processing applications.

Steps involved in Pixel Attribution Methods:

Step 1. Get images dataset and preprocess it.
Step 2. Apply a DL model on it.
Step 3. Get the model's prediction output.
Step 4. Apply Pixel Attribution method on the model for a test image and get the pixel-wise importance value.
Step 5. A heatmap is over imposed on the original image to showcase which pixels are triggered by the model in getting the prediction outcome.

Types of Pixel Attribution Methods

1. Gradient-based methods – Saliency Maps and Integrated Gradients are of this type
2. Backpropagation-based methods – Guided Backpropagation method
3. CAM Methods – GradCAM, GradCAM++
4. Perturbation methods – Occlusion method and RISE method

In the below section, we discuss GradCAM visualization with sample output shown in Figure 4.3.

Sample code for GradCAM

```
from sklearn.model_selection import train_test_split

X_train, X_test, y_train, y_test  = train_test_
   split(res['image_path'], target, test_size=0.33,
   random_state=42)
print(f"train shape: {X_train.shape}- y_train shape: {y_train.
   shape}")
print(f"test shape: {X_test.shape}- y_test shape: {y_test.
   shape}")
fig, axis = plt.subplots(3, 2, figsize=(30, 30))
for images, ax in zip(next(iter(test_loader))[0][:6], axis.
   flat):
    img_array = get_img_array(images)
    # Remove last layer's softmax
    vit_classifier.layers[-1].activation = None
    # Print what the top predicted class is
    preds = vit_classifier.predict(img_array)
    heatmap = gradcam_heatmap(img_array, vit_classifier,
       last_conv_layer_name)

    heatmap = np.reshape(heatmap, (36,36))
    display_gradcam(images, heatmap, preds=preds[0], plot=ax)
```

4.5 PROTOTYPE-BASED EXPLANATIONS

The prototype-based explanations explain a model's prediction based on the comparison results of input instance to representative examples from the training dataset. The prototypes are identified after model training without changing the actual DL model.

Steps involved in prototype-based explanations

Step 1. Select the dataset and preprocess it.
Step 2. Train a DL model on the dataset.
Step 3. Identify some representative prototypes.
Step 4. Compare the input image to the prototypes.
Step 5. Based on the similarity, generate explanations.

4.5.1 ProtoPNet – A Prototype-Based Explanation Model

An antehoc explainability model called Prototype-based Networks (ProtoPNet), which use representative examples or "prototypes" to explain decisions. Additionally, we implemented Interactive Machine Learning (IML) in combination with ProtoPNet to further refine model explanations with human input. This approach provides an even deeper level of transparency, allowing both the model and the clinician to interact, thus enhancing the understanding of the predictions.

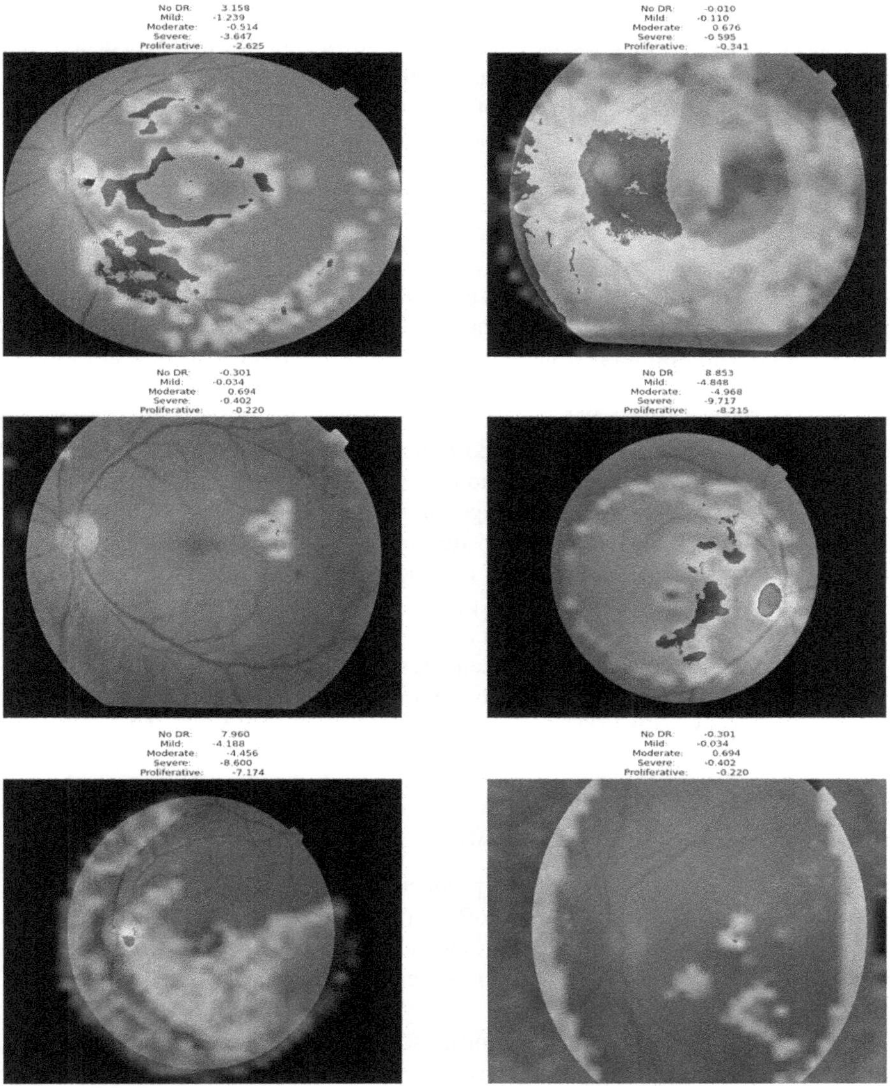

FIGURE 4.4 GradCAM visualization of images with feature importance highlighted.

ProtoPNet explains predictions by comparing parts of the test images with learned "prototypes" of image patches from the training set. For instance, it can show similar regions of the brain from past MRI images that were labeled as stroke, providing clinicians with concrete visual evidence.

The combination of IML and ProtoPNet integrates the interpretable and transparent aspects of IML, which focuses on breaking down complex model decisions into human-understandable reasoning, with ProtoPNet's ability to provide visual evidence. This hybrid approach offers both visual and textual explanations, making the decision-making process more transparent to clinicians.

4.5.2 ProtoPNet Explanations

ProtoPNet produces explanations by comparing input images to learned prototype vectors, allowing the model to provide interpretable insights into its decision-making process. The core idea is that the model learns a set of representative prototypes during training, which correspond to critical patterns or features from the dataset.

During inference, ProtoPNet computes distances between the features extracted from the input image and these learned prototypes. The closest prototypes are then used to inform the classification decision.

By visualizing these closest prototypes alongside the input images, ProtoPNet can highlight specific regions or characteristics in the input that led to the model's decision. This approach makes the model's predictions more interpretable, as clinicians or users can see which learned prototype (i.e., medical feature or pattern) the model used to classify an image as either "Ischemic" or "Control."

This method bridges the gap between DL's high accuracy and the need for transparency in clinical decision-making by showing how specific patterns in the data match learned prototypes. The visualizations using this XAI model is given in Figures 4.5 and 4.6.

4.5.3 IML + ProtoPNet Explanations

The IML combined with ProtoPNet produces a more dynamic and user-centered approach to explanations. While ProtoPNet offers prototype-based explanations by identifying the most similar learned prototype to an input image, the addition of IML introduces an interactive layer where users can provide real-time feedback on model predictions.

After the model makes its predictions, users are prompted to assess whether the prediction is correct. If it is incorrect, users can provide the correct label, allowing for direct interaction and the potential to collect data for model improvement.

This interaction gives users a better understanding of the model's behavior, while also contributing to its continual learning.

The combination of IML with ProtoPNet not only highlights the reasons behind the model's decision (via prototypes) but also provides a means to refine the model with human feedback, enhancing both interpretability and accuracy over time. This approach is particularly valuable in sensitive domains like medical imaging, where expert feedback can lead to more trustworthy and fine-tuned AI models. It is given in Figures 4.6 and 4.7.

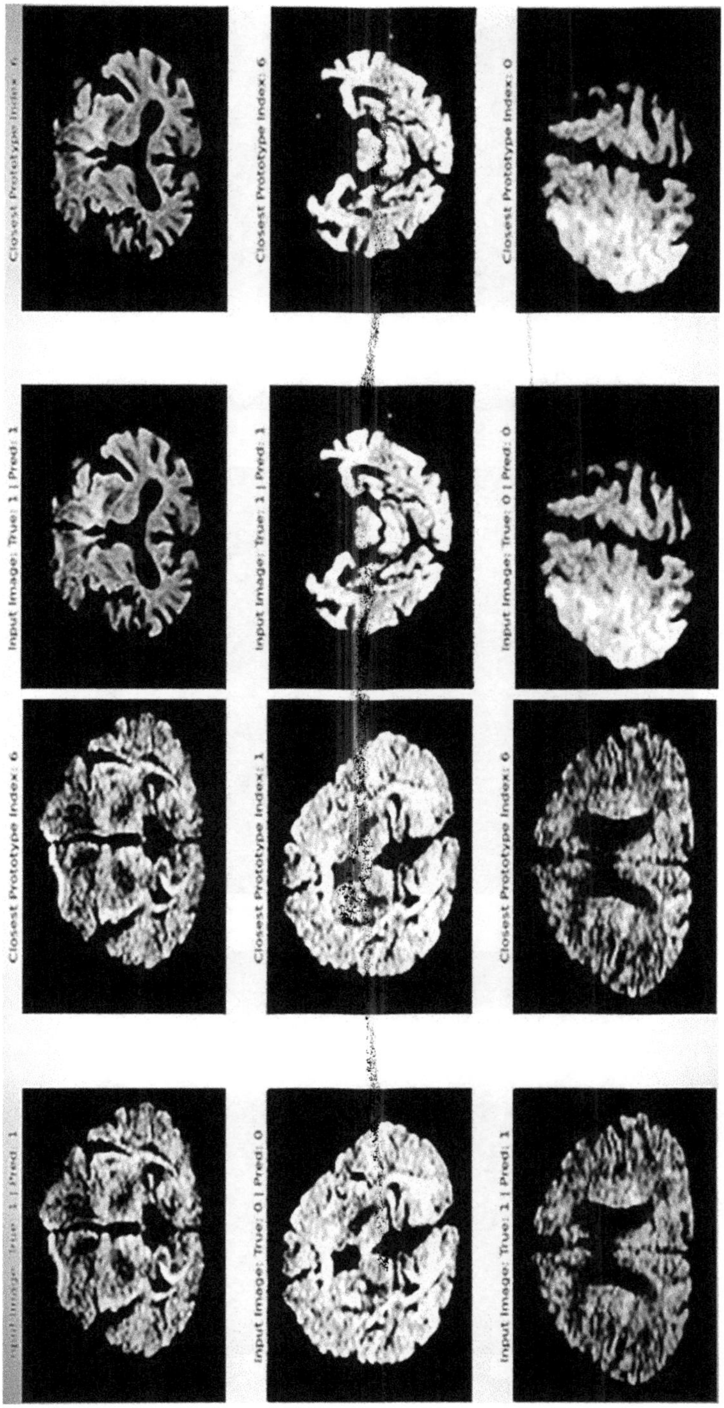

FIGURE 4.5 ProtoPNet explanations.

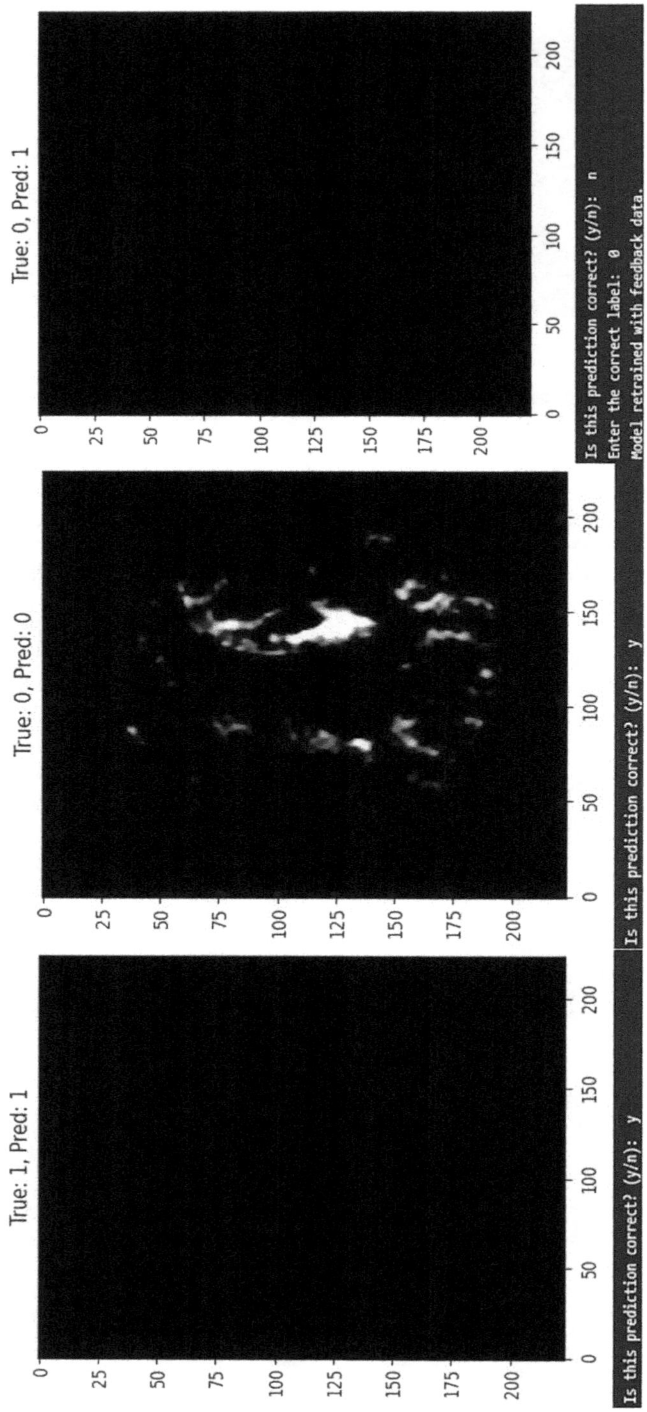

FIGURE 4.6 IML + ProtoPNet training.

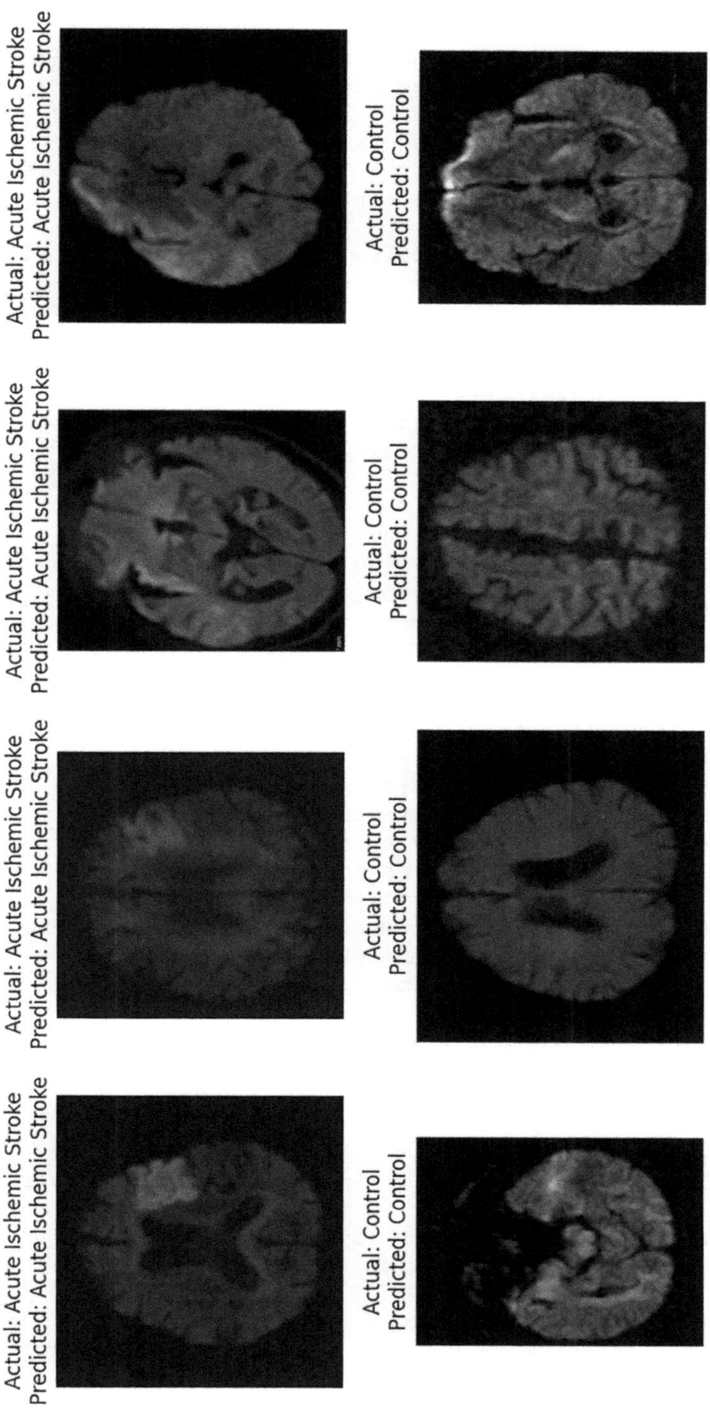

FIGURE 4.7 IML with ProtoPNet visualization.

Sample Code:

```
class ProtoPNet(nn.Module):
    def __init__(self, base_model, num_prototypes, num_
            classes, prototype_dim=512):
        super(ProtoPNet, self).__init__()
        # Use layers up to the last convolutional block
        self.feature_extractor = nn.Sequential(*list(base_
           model.children())[:-2])

        # Adaptive average pooling layer to ensure fixed
           output size
        self.pool = nn.AdaptiveAvgPool2d((1, 1))  # Output
           size of 1x1

        # Project features to the prototype dimension
        self.feature_projection = nn.Linear(2048, prototype_
           dim)  # Assuming ResNet50's output is 2048

        # Prototype layer
        self.num_prototypes = num_prototypes
        self.prototype_layer = nn.Parameter(torch.randn(num_
           prototypes, prototype_dim))

        # Fully connected layer for final classification
        self.fc = nn.Linear(num_prototypes, num_classes)
    def forward(self, x):

        # Extract features
        features = self.feature_extractor(x)
        features = self.pool(features)
        features = features.view(features.size(0), -1)  #
           Flatten the features to (batch_size, 2048)

        # Project features to match the prototype dimension
        features = self.feature_projection(features)  # Now
           shape will be (batch_size, prototype_dim)

        # Compute distances to prototypes
        distances = torch.cdist(features.unsqueeze(0), self.
           prototype_layer.unsqueeze(0)).squeeze(0)

        # Compute logits
        logits = -distances  # Minimize distance to correct
           class prototypes
        # Final classification layer
        output = self.fc(logits)

        return output, distances
# Define the ResNet model as a base model (you can change it
   to ResNet50, ResNet101, etc.)
base_model = ResNet152(num_classes=2).to(device)
```

```
num_classes = 2  # Number of classes for your task (e.g.,
   Stroke and Control)
num_prototypes = 10  # Number of prototypes you want to learn
prototype_dim = 512  # Dimension of the prototype (adjust
   based on your feature size)

# Initialize the ProtoPNet model
proto_model = ProtoPNet(base_model, num_prototypes=
   num_prototypes, num_classes=num_classes,
   prototype_dim=prototype_dim)
proto_model = model.to(device)
# Training settings
criterion = nn.CrossEntropyLoss()
optimizer = optim.SGD(model.parameters(), lr=0.01,
   momentum=0.9, weight_decay=0.0001)
scheduler = optim.lr_scheduler.ReduceLROnPlateau(optimizer,
   factor=0.1, patience=5)
epochs = 100
# Displaying the results
plt.figure(figsize=(12, 6))

for i in range(8):
    plt.subplot(2, 4, i + 1)
    plt.imshow(images[i])
    plt.title(f'Actual: {actual_labels[i]}\nPredicted:
       {predicted_labels[i]}')
    plt.axis('off')

plt.tight_layout()
plt.show()
```

4.6 COUNTERFACTUAL EXPLANATIONS

The counterfactual explanations method helps to explain a model's prediction by means of generating alternative scenarios. A typical scenario is: if a slight change is made in the feature's value, what will be its impact on the output? This is generally applied after model training. It focuses on actionable changes in the input.

Types of Counterfactual Explanations

1. Feature-based – Changes are applied in feature values.
2. Actionable features-based – Changes are applied only in the output-controllable features.
3. Diverse method – More than one counterfactual explanation is applied.
4. Plausible method – Uses data distribution as the primary factor.

4.6.1 DiCE Method – Counterfactual Explanation Method

DiCE is expanded as Diverse Counterfactual Explanations method, which is a posthoc explainability method that generates diverse counterfactual explanations for a given prediction.

Some of the key components used by DiCE are: validity loss, proximity loss, diversity term, and constraints to generate counterfactuals.

Code:
```
dice_data = dice_ml.Data(
    dataframe=df,
    continuous_features=X.columns.tolist(),
    outcome_name='target'
)
dice_model = dice_ml.Model(
    model=model,
    backend="sklearn"
)
explainer = dice_ml.Dice(
    dice_data,
    dice_model,
    method="random"
)
query_instance = X_test.iloc[[0]]
counterfactuals = explainer.generate_counterfactuals(
    query_instance,
    total_CFs=2,
    desired_class="opposite"
)
counterfactuals.visualize_as_dataframe()
```

Output from DiCE for "dataset = helpers.load_adult_income_dataset()" is given in Tables 4.1 and 4.2.

TABLE 4.1
Original Feature Values of a Sample Data

Age	Workclass	Education	marital_ status	Occupation	Race	Gender	Hours_ per_ week	Income
29	Private	HS-grad	Married	Blue-Collar	White	Female	38	0

TABLE 4.2
Diverse Counterfactual Set (New Outcome: 1)

Counter factual #	Age	Workclass	Education	Marital_ status	Occupation	Race	Gender	Hours_ per_ week	Income
0	29	Private	HS-grad	Married	Professional	White	Female	59	1
1	29	Private	Some-college	Married	Blue-Collar	White	Female	97	1
2	36	Private	Assoc	Married	Blue-Collar	White	Female	38	1
3	40	Private	HS-grad	Married	White-Collar	White	Female	38	1

4.7 SUMMARY

As DL models are inherently black box in nature, they need to prove their trustworthiness before deployment. To do this, posthoc explainability methods are applied once the DL models provide the prediction. Some of the posthoc explainability methods applied include LIME, SHAP, and Saliency Maps to identify the important features that decide the model's outcome.

BIBLIOGRAPHY

Abbaoui, W., Retal, S., Ziti, S., El Bhiri, B., and Moussif, H. (2024). Ischemic stroke classification using VGG-16 convolutional neural networks: A study on Moroccan MRI scans. *International Journal of Online and Biomedical Engineering*, 20(2):6.

Angelov, P., Soares, E., Jiang, R., Arnold, N., and Atkinson, P. (2021). Explainable artificial intelligence: An analytical review. *WIREs Data Mining and Knowledge Discovery*, 11(5). https://doi.org/10.1002/widm.1424

Athisayamani, S., Antonyswamy, R. S., Sarveshwaran, V., Almeshari, M., Alzamil, Y., and Ravi, V. (2023). Feature extraction using a residual deep convolutional neural network (ResNet-152) and optimized feature dimension reduction for MRI brain tumor classification. *Diagnostics*, 13(4):668.

Cao, Y., Li, S., Liu, Y., Yan, Z., Dai, Y., Yu, P. S., and Sun, L. (2023). A comprehensive survey of AI-generated content (AIGC): A history of generative AI from GAN to ChatGPT. arXiv abs/2303.04226.

Chae, M., Gil, H.-W., Cho, N.-J., and Lee, H. (2022). Machine learning-based cardiac arrest prediction for early warning system. *Mathematics*, 10:2049. https://doi.org/10.3390/math10122049

Dave, Devam, Naik, Het, Singhal, Smiti, and Patel, Pankesh. (2020). Explainable AI meets healthcare: A study on heart disease dataset. arXiv preprint arXiv:2011.03195.

Farrag, A., Gad, G., Fadlullah, Z. M., Fouda, M. M., and Alsabaan, M. (2023). An Explainable AI system for medical image segmentation with preserved local resolution: Mammogram tumor segmentation. *IEEE*, 11:125543–125561.

Gabbay, F., Bar-Lev, S., Montano, O., and Hadad, N. (2021). A LIME-based explainable machine learning model for predicting the severity level of COVID- 19 diagnosed patients. *Applied Sciences*, 11(21):10417.

Hossain, Shahriar, Chakrabarty, Amitabha, Gadekallu, Thippa Reddy, Alazab, Mamoun, and Piran, Md Jalil. (2024). Vision transformers, ensemble model, and transfer learning leveraging Explainable AI for brain tumor detection and classification. *IEEE Journal of Biomedical and Health Informatics*, 28(3):1261–1272.

https://www.kaggle.com/datasets/fedesoriano/heart-failure-prediction

Ivanovic, M., and Balaz, I. (2020). Influence of artificial intelligence on personalized medical predictions, interventions and quality of life issues. In *24th International Conference on System Theory, Control and Computing (ICSTCC)*, Sinaia, Romania, pp. 445–450. https://doi.org/10.1109/ICSTCC50638.2020.9259674

Ivanovic, Mirjana, and Macos, Dragan. (2023). Applications of artificial intelligence methods in medicine. In *International Conference on Information Technology (ICIT)*, Amman, Jordan, pp. 349–354, IEEE. https://doi.org/10.1109/ICIT58056.2023.10226107

Jee, S. H., Jang, Y., Oh, D. J., Oh, B. H., Lee, S. H., Park, S. W., Seung, K. B., Mok, Y., Jung, K. J., Kimm, H., Yun, Y. D., Baek, S. J., Lee, D. C., Choi, S. H., Kim, M. J., Sung, J., Cho, B., Kim, E. S., Yu, B. Y., Lee, T. Y., Kim, J. S., Lee, Y.-J., Oh, J.-K., Kim, S. H., Park, J.-K., Koh, S. B., Park, S. B., Lee, S. Y., Yoo, C.-I., Kim, M. C., Kim, H.-K., Park, J.-S., Kim, H. C., Lee, G. J., and Woodward, M. (2014). A coronary heart disease prediction model: The Korean heart study. *BMJ Open*, 4(5):e005025. https://doi.org/10.1136/bmjopen-2014-005025

Junsheng, M., Kadoch, M., Yuan, T., Lv, W., Liu, Q., and Li, B. (2024). Explainable federated medical image analysis through causal learning and Blockchain. *IEEE Journal of Biomedical and Health Informatics*, 28(6):3206–3218.

La Salvia, Marco, Torti, Emanuel, Leon, Raque, Fabelo, Hima, Ortega, Samuel, Martinez-Vega, Beatri, Callico, Gustavo M., and Leporati, Francesc. (2022). Deep convolutional generative adversarial networks to enhance artificial intelligence in healthcare: A skin cancer application. *Sensors*, 22(16):6145. https://doi.org/10.3390/s22166145

Lauritsen, S. M., Kristensen, M., Olsen, M. V., Larsen, M. S., Lauritsen, K. M., Jørgensen, M. J., Lange, J., and Thiesson, B. (2020). Explainable artificial intelligence model to predict acute critical illness from electronic health records. *Nature Communications*, 11:3852.

Lee, H., Yang, H. L., Ryu, H. G., Jung, C. W., Cho, Y. J., Yoon, S. B., Yoon, H. K., and Lee, H. C. (2023). Real-time machine learning model to predict in- hospital cardiac arrest using heart rate variability in ICU. *NPJ Digital Medicine*, 6:215. https://doi.org/10.1038/s41746-023-00960-2

Loh, Hui We, Ooi, Chui Pin, Seoni, Silvi, Barua, Prabal Datt, Molinari, Filipp, and Acharya, U. Rajendr. (2022). Application of explainable artificial intelligence for healthcare: A systematic review of the last decade (2011–2022). *Computer Methods and Programs in Biomedicine*, 226:107161.

Loh, Hui Wen, Yarahmadi, A., Hsu, C. C., Biyari, M., Sookhak, M., Ameri, R., Dehzangi, I., Chronopoulos, A. T., and Liang, H. W. (2022). Application of explainable artificial intelligence in medical health: A systematic review of interpretability methods. *Informatics in Medicine Unlocked*, 40:101286.

Mahim, S. M., Ali, M. S., Hasan, M. O., Nafi, A. A. N., Sadat, A., Al Hasan, S., Shareef, B., Ahsan, M. M., Islam, M. K., Miah, M. S., and Niu, M. B. (2024). Unlocking the potential of XAI for improved Alzheimer's disease detection and classification using a ViT-GRU model. *IEEE*, 12:8390–8412.

Mao, Xuewei, Shan, W., and Yu, J. (2024). Automatic diagnosis and subtyping of ischemic stroke based on a multidimensional deep learning system. *IEEE Transactions on Instrumentation and Measurement*, 73:1–11.

Mung, P. S., and Phyu, S. (2020). Effective analytics on healthcare big data using ensemble learning. In *IEEE Conference on Computer Applications (ICCA)*, Yangon, Myanmar, pp. 1–4. https://doi.org/10.1109/ICCA49400.2020.9022853

Nag, Anindya, Mondal, Hirak, Hassan, Md Mehedi, Al-Shehari, Taher, Kadrie, Mohammed, Al-Razgan, Muna, Alfakih, Taha, Biswas, Sujit, and Bairagi, Anupam Kumar. (2024). TumorGANet: A transfer learning and generative adversarial network based data augmentation model for brain tumor classification. *IEEE Access*, 12:103060–103081.

Nova, Kannan. (2023). Generative AI in healthcare: Advancements in electronic health records, facilitating medical languages, and personalized patient care. *Journal of Advanced Analytics in Healthcare Management*, 7(1):115–131. https://research.tensorgate.org/index.php/JAAHM/article/view/43

Palkar, Anisha. (2024). Empowering glioma prognosis with transparent machine learning and interpretative insights using Explainable AI. *IEEE Access*, 12:31697–31718.

Parimala, Nallamot, and Muneeswari, G. (2024). Modified ResNet152v2: Binary classification and hybrid segmentation of brain stroke using transfer learning-based approach. *Polish Journal of Medical Physics and Engineering*, 30(1):24–35.

Reddy, G. P., and Kumar, Y. V. P. (2023). Explainable AI (XAI): Explained. In *IEEE Open Conference of Electrical, Electronic and Information Sciences (eStream)*, Vilnius, Lithuania, pp. 1–6. https://doi.org/10.1109/eStream59056.2023.10134984

Sakri, Sapiah, Basheer, Shakila, Zain, Zuhaira Muhammad, Ismail, Nurul Halimatul Asmak, Nassar, Dua Abdellatef, Aldehim, Ghadah Nasser, and Alharaki, Mais Ayman. (2024). An improved concatenation of deep learning models for predicting and interpreting ischemic stroke. *IEEE Access*, 12:53189–53204.

Saraswat, D., Bhattacharya, P., Verma, A., Prasad, V. K., Tanwar, S., Sharma, G., Bokoro, P. N., and Sharma, R. (2022). Explainable AI for healthcare 5.0: Opportunities and challenges. *IEEE Access*, 10:84486–84517. https://doi.org/10.1109/ACCESS.2022.3197671

Sharma, Y., Verma, A., Rao, K., and Eluri, V. (2020). "Reasonable explainability" for regulating AI in health. *ORF Issue Brief*, 401:1–16.

Shokrollahi, Yasin, Yarmohammadtoosky, Saha, Nikahd, Matthew M., Dong, Pengfe, Li, Xianqi, and Linxia, Gu. (2023). A comprehensive review of generative AI in healthcare. arXiv abs/2310.00795.

Singh, Amitojdeep, Sengupta, Sourya, and Lakshminarayanan, Vasudevan. (2020). Explainable deep learning models in medical image analysis. *Journal of Imaging*, 6(6):52.

Stilou, S., Bamidis, P., Maglaveras, N., and Pappas, C. (2001). Mining association rules from clinical databases: An intelligent diagnostic process in healthcare. *Studies in Health Technology and Informatics*, 84:1399–1403.

Suryadevara, Chaitanya. (2020). Towards personalized healthcare – An intelligent medication recommendation system. *IEJRD – International Multidisciplinary Journal*, 5:1–16.

Wang, W., Lin, X., Feng, F., He, X., and Chua, T. (2023). Generative recommendation: Towards next-generation recommender paradigm. arXiv, abs/2304.03516.

Wiens, J., and Shenoy, E. S. (2018). Machine learning for healthcare: On the verge of a major shift in healthcare epidemiology. *Clinical Infectious Diseases: An Official Publication of the Infectious Diseases Society of America*, 66(1):149–153. https://doi.org/10.1093/cid/cix731

Yang, C. C. (2022). Explainable artificial intelligence for predictive modeling in healthcare. *Journal of Healthcare Informatics Research*, 6(2):228–239. https://doi.org/10.1007/s41666-022-00114-1

Yue, Wenbin, Wang, Zidong, Zhang, Jieyu, and Liu, Xiaohui. (2021). An overview of recommendation techniques and their applications in healthcare. *IEEE/CAA Journal of Automatica Sinica*, 8(4):701–717. https://doi.org/10.1109/JAS.2021.1003919

Zhang, Peng, and Kamel Boulos, Maged N. (2023). Generative AI in medicine and healthcare: Promises, opportunities and challenges. *Future Internet*, 15(9):286. https://doi.org/10.3390/fi15090286

Zhang, Hongjia, and Ogasawara, Katsuhik. (2023). Grad-CAM-based explainable artificial intelligence related to medical text processing. *Bioengineering*, 10(9):1070.

Part 2

Integration of XAI into DLOps

5 DLOps
Introduction to XAI in DL Model Development

5.1 INTRODUCTION

5.1.1 Generic Workflow of XAI

The generic workflow of XAI is shown in Figure 5.1. The training data taken from the given dataset is used by the ML/DL model to get trained. Using test data, the model is evaluated, and the predictions are derived from the model. If the model's outcome is of acceptable accuracy level, then the model can be accepted for deployment. Otherwise, the model is fine-tuned to improve the accuracy.

When explainability is incorporated, the model will now be passed to another process – XAI method such as LIME/SHAP etc. with the training data and the test data taken from the test dataset. Based on the outcome, the XAI method will give the appropriate explanations to the end users so that the model can be understood on its working mechanism.

5.1.2 Integrated XAI Framework

The framework introduced here comprises both global and local interpretability. The global interpretability focuses on the model's final outcome, which is intrinsically interpretable as far as simple ML models are concerned.

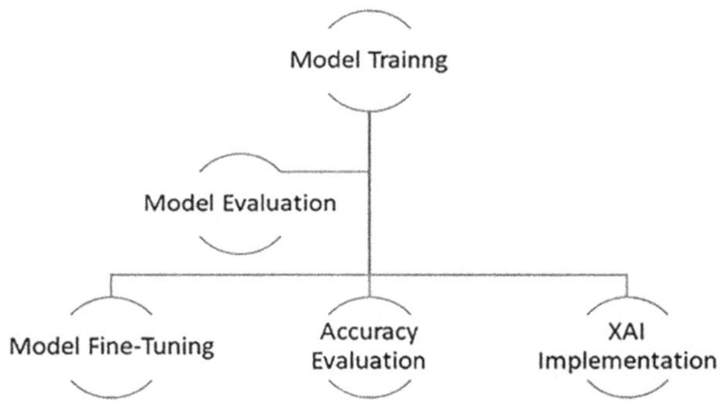

FIGURE 5.1 Generic workflow of XAI.

DOI: 10.1201/9781003630241-7

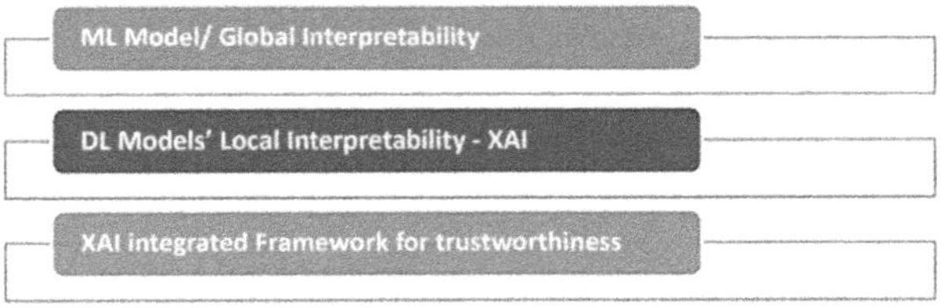

FIGURE 5.2 XAI integrated framework with local and global interpretability.

Whereas for complex DL models, each neuron-wise, layer-wise interpretation is essential as it is highly complex to interpret any internal working based on the final outcome alone.

The proposed framework is shown in Figure 5.2 for basic level understanding.

In order to provide explanations to the end users, the XAI method should contain the following properties: it must take input, produce output, support interactivity with end users, and include any other formal constraints required for the explanation required from the model.

One can consider an XAI method as simply as an implemented function that outputs explanations. This allows structuring aspects of the XAI method into the main defining aspects of a function: the input, the output, the function class described by mathematical properties or constraints, and the actual processing of the explanation function like its interactivity.

5.2 DLOPS – AN INTRODUCTION

DLOps is a set of practices, tools, and workflows that automate, monitor, version, deploy, and maintain the DL models.

DLOps addresses the following challenges generally faced by traditional DL model development:

- Large Dataset – DLOps addresses it by data versioning and lineage.
- GPU/TPU dependency for Computations – DLOps helps to achieve it through resource scheduling.
- Long training time – DLOps provides automated pipelines to address this.
- Model Drifts – The continuous monitoring process of DLOps will help in addressing it.
- Black-box model – Application of XAI in the DLOps development cycle will help understand the internal workings of the model.
- Retraining repetition – Applying CI/CD in the DL development leads to addressing this.
- Reproducibility issues – Tracking of experiments will find the issues in reproducibility of the model to other experiments.

5.2.1 DLOps Platforms

Some of the cloud-based DLOps platforms are listed below:

- AWS – SageMaker, SageMaker Pipelines, SageMaker Clarify (XAI), and CloudWatch
- Azure – Azure ML Studio, Azure DLOps, and Responsible AI Dashboard
- Google Cloud – Vertex AI, TFX, and Explainable AI

XAI Integration in DLOps

- As DL models are providing high accuracy with low interpretability and explainability of the outcome due to their black-box nature, XAI methods are applied in high-risk domains such as healthcare, finance, industrial automation, self-driving cars, etc.
- XAI integration in DLOps ensures transparency, trust, compliance, and continuous monitoring and governance of the entire DL model life cycle development.

5.2.2 Laws for Explainability

As per GDPR law, it is essential to provide explanations on why and how a model has taken a critical decision. Some of the acts such as AI Act (EU), healthcare regulations such as FDA, HIPAA, and other financial-based regulations are also expecting reliable DL model development as AI has increasingly become a part of everyday applications.

5.2.3 DLOps Life Cycle

The major life cycle phases of DLOps are given in Figure 5.3.

5.2.4 DLOps Stages with XAI

Each of the DLOps development life cycle stages will be associated with the role of XAI to meet the above requirements.

- DLOps stage: Data Ingestion ➔ XAI's role is feature attribution and bias detection
- DLOps stage: Model Training ➔ XAI's role is explanation-aware model selection
- DLOps stage: Validation ➔ XAI's role is trust and fairness evaluation
- DLOps stage: Deployment ➔ Explainability APIs of XAI
- DLOps stage: Monitoring Stage ➔ Drift-based explanation changes
- DLOps stage: Retraining Stage ➔ Explanation-guided feedback

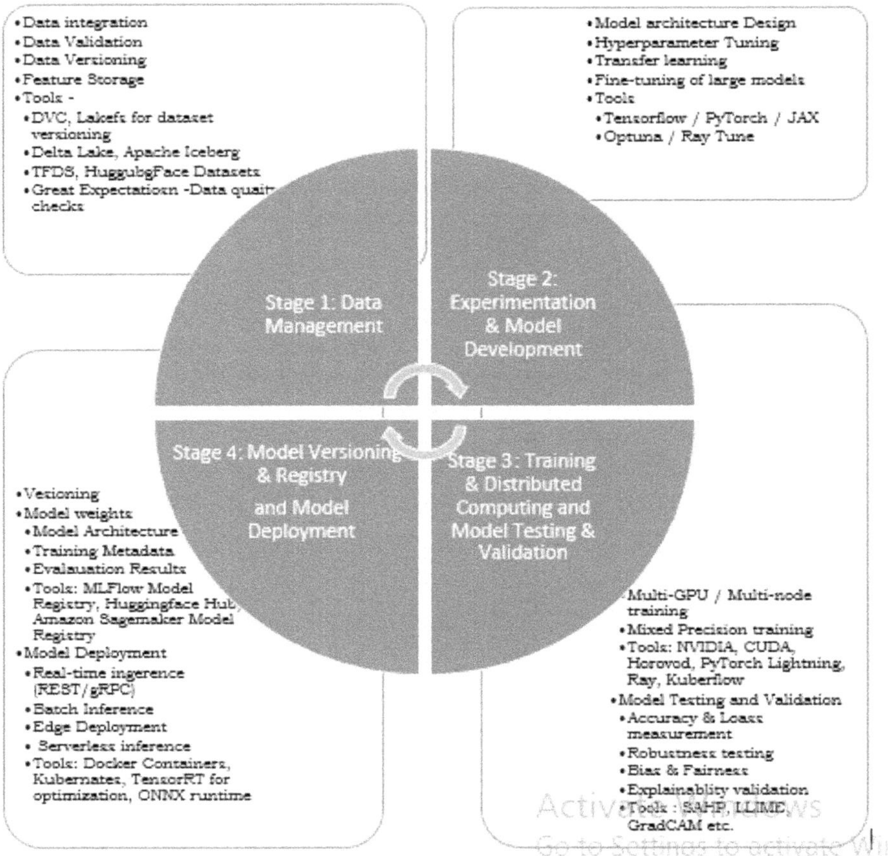

FIGURE 5.3 DLOps life cycle with XAI.

The XAI integrated DLOps framework is given in Figure 5.4.

5.3 XAI IN DLOPS STAGES

The role of XAI in each stage of DLOps is given in detail as below.

5.3.1 XAI in Data Layer

The primary objectives at this layer are to identify biased features, detect spurious correlations, and understand feature distributions.

Some of the techniques applied at this layer are: statistical feature importance analysis, SHAP summary plots, correlation heatmaps, and data slices analysis. The XAI methods applied are: SHAP, Pandas Profiling, IBM AI Fairness 360.

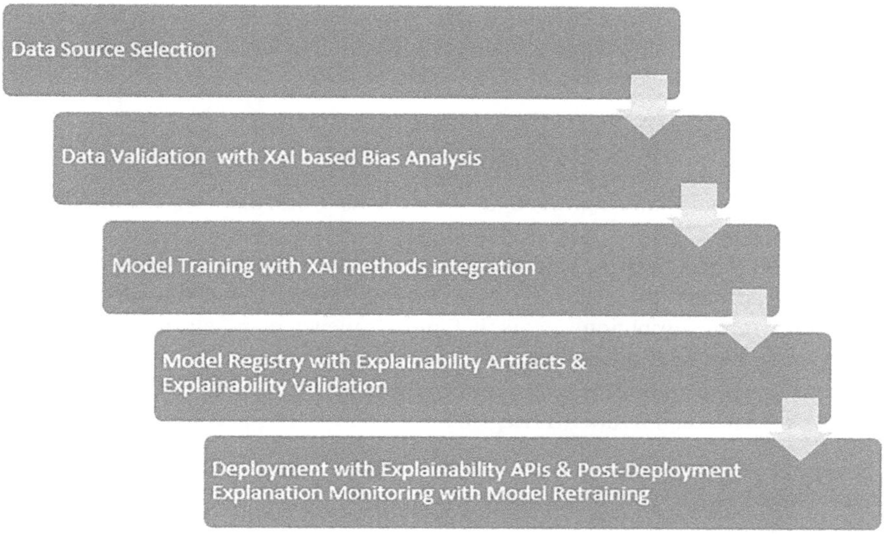

FIGURE 5.4 XAI integrated DLOps stages.

5.3.2 XAI in Model Training Layer

The XAI methods applied on different DL models are given below:

- CNN model – GradCAM
- RNN/LSTM – Attention Validation
- Transformers – Attention Rollout
- Other DL models – SHAP, Integrated Gradients, etc.

The XAI integration in these models should be stored as training artifacts. The method applied will compare the SHAP values across epochs, and reject models without stable explanations across epochs.

Some of the tools that can be used are: Captum from PyTorch, TF Explain, and SHAP Deep Explainer.

5.3.3 XAI Aware Model Validation Layer

Some of the most important dimensions of validation layer are supported by XAI. The accuracy is provided with explanations on misclassifications so that the developer can identify the issues in the model and fine-tune it appropriately. The fairness evaluation is done by performing group-wise comparison using XAI methods. To evaluate robustness of the model, the consistency of the explanations generated will be used. The trustworthiness of the DL model can be accompanied by human-in-the-loop validation so that experts' judgment can be used to evaluate the reliability of the model.

Some of the evaluation metrics associated with validation layer are: explanation stability, feature attribution variance, faithfulness, fidelity, interpretability size, and sparsity.

5.3.4 Explainability Artifacts in Model Registry Layer

When the DL model is developed, all the associated artifacts such as dataset, model parameters, model architecture, and results associated with that version will be stored in the repository.

However, when XAI is integrated in the DLOps cycle, apart from the above-mentioned artifacts, the outcomes such as SHAP plots, LIME output, GradCAM heatmaps, feature attribution vectors, ICE plots, saliency maps, etc. must be stored in the repository for later referencing.

Some of the useful tools associated with the above include: MLflow, Hugging Face Model Hub, and Azure ML Registry.

5.3.5 XAI at Deployment Layer

It is essential to decide whether we need to have XAI as an embedded part of the DL model or as an On-Demand one based on the need.

If it is decided to have the XAI methods embedded in the developed DL system as per the client's requirement, then explanations will be generated along with the prediction results. It has higher latency when compared to the normal DL system without XAI embedding. However, it is highly essential to have it in modern healthcare applications, finance applications, and any other critical applications that need to satisfy the responsiveness and ethical aspects of the DL model.

If the decision is to apply XAI methods on an on-demand basis, then the developers can have a dashboard with an option to have XAI in order to generate explanations along with the predictions only when there is a requirement. It is highly preferable in low-powered edge devices, large-scale systems in which so many intermittent predictions will be generated by the model.

5.3.6 XAI-Based Post Deployment Explanation Monitoring

The reason behind having XAI-based post deployment monitoring is to check the model when the accuracy is high, but reasoning may drift and there is an emergence of hidden bias. This monitoring process will be enabled when there is a drift in feature importance, explanation divergence, or input distribution shift.

Some of the tools used for such monitoring are: Evidently AI, Fiddler, etc.

5.3.7 XAI Driven Continuous Retraining

The DL model needs to be retrained if the explanation drift threshold is exceeded or there is a degradation in bias metrics or there is a human feedback on the wrong predictions of the model.

Generally, a feedback loop will be added in which if there is a change in explanation, then human-in-the-loop-based evaluation will be done and then if there is a need, the dataset will be updated. Finally, the model will be retrained on the updated dataset.

5.3.8 CI/CD with XAI

During Continuous Integration (CI) pipeline of DLOps, the explanation consistency will be evaluated and the final build will fail if the bias threshold is exceeded.

In Continuous Delivery (CD) pipeline, the explanations generated during intermittent deployments will be evaluated for their consistency and if there is an issue, then the model will be updated accordingly. To do explainability testing, some dummy models of the actual model will be created to evaluate the trustworthiness of the explanations generated.

5.4 XAI INTEGRATED DL MODEL DEVELOPMENT FRAMEWORK

Phase I. Design a Deep Learning model for the given problem specification.
Phase II. Apply XAI methods to generate explanations on the model trained and tested before deployment.
Phase III. Evaluate the XAI methods using fidelity, unambiguity, and interpretability size metrics.

The entire model framework is given in Figure 5.5.

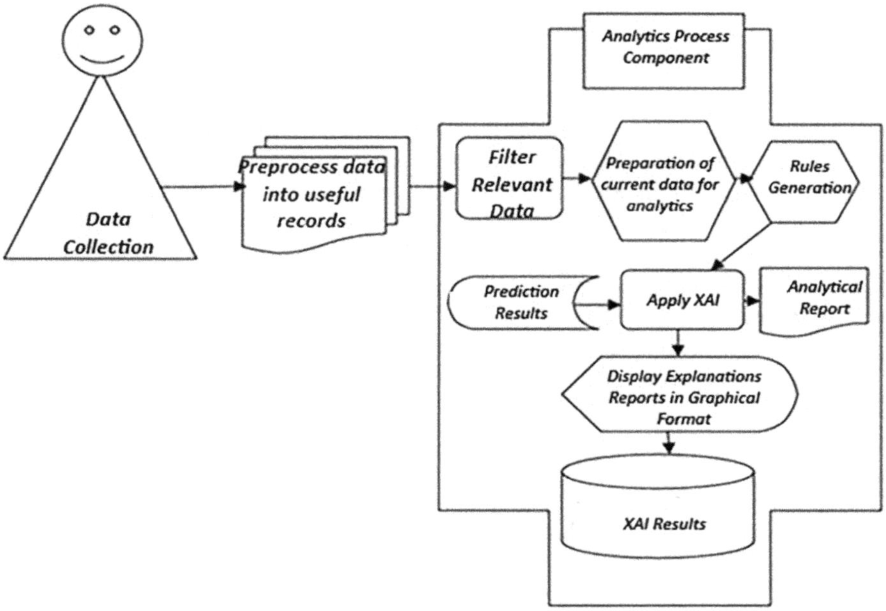

FIGURE 5.5 XAI framework for explanations.

The proposed XAI framework has the following steps in it.

Step 1. Extract patient data using the assessment by means of humans or wearable devices.
Step 2. Preprocess the data in order to use it for further data analytics.
Step 3. Enter into the Analytics component which has the following functionalities:
  - Step 3.1. Filter only relevant data for further processing.
  - Step 3.2. Apply feature engineering to fine-tune the data and the features to make it ready for analytics.
  - Step 3.3. Generate rules to define the methodology to be followed in arriving at the prediction.
  - Step 3.4. Get the predictions from the model for the given dataset.
  - Step 3.5. Apply XAI methods on the developed ML/DL model to generate explanations for the prediction.
  - Step 3.6. Generate visualizations and reports from the XAI methods.
  - Step 3.7. Validate the ML/DL model's outcome from the XAI method's output.
  - Step 3.8. Retrain the model if it is not providing the decision as per the given rules, and repeat from steps 3.4 to 3.8.
  - Step 3.9. If the results are satisfactory, then evaluate the XAI method by means of various metrics and measures such as fidelity, unambiguity, and predictability size.

Step 4. Store the visualizations and the model for further data analytics.

5.5 SUMMARY

XAI is playing a vital role in the trustworthiness assessment of DL models. Nowadays, the XAI methods are integrated into the DL model development at each and every stage of development. This increases transparency, trustworthiness, and debugging efficiency of the DL model development. It also ensures the model's behavior alignment with human-in-the-loop and domain knowledge. During CI, the explanation consistency will be evaluated and the bias-based analysis will be conducted to confirm the model for deployment. In CD, the explanations are generated during each and every intermittent delivery to ensure consistency in outcome. If there are any issues, then the model will be appropriately updated and evaluation metrics are used to ensure the various product metrics such as feature attribution variances, fidelity, interpretability size, and sparsity.

BIBLIOGRAPHY

Amou Najafabadi, F., Bogner, J., Gerostathopoulos, I., and Lago, P. (2024, September). An analysis of MLOps architectures: A systematic mapping study. In *European Conference on Software Architecture*, Luxembourg (pp. 69–85). Cham: Springer Nature Switzerland.

Avacharmal, R., and Pamulaparthyvenkata, S. (2022). Enhancing algorithmic efficacy: A comprehensive exploration of machine learning model lifecycle management from inception to operationalization. *Distributed Learning and Broad Applications in Scientific Research*, 8, 29–45.

Bayram, F., and Ahmed, B. S. (2025). Towards trustworthy machine learning in production: An overview of the robustness in MLOps approach. *ACM Computing Surveys*, 57(5), 1–35.

Eken, B., Pallewatta, S., Tran, N., Tosun, A., and Babar, M. A. (2025). A multivocal review of MLOps practices, challenges and open issues. *ACM Computing Surveys*, 58(2), 1–35.

Faubel, L., Woudsma, T., Methnani, L., Ghezeljhemeidan, A. G., Buelow, F., Schmid, K., van Driel, W. D., Kloepper, B., Theodorou, A., Nosratinia, M., and Bång, M. (2023). Towards an MLOps architecture for XAI in industrial applications. arXiv preprint arXiv:2309.12756.

Jain, N. S. (2023). Integrating artificial intelligence with DevOps: Enhancing continuous delivery, automation, and predictive analytics for high-performance software engineering. *World Journal of Advanced Research and Reviews*, 17, 1025–1043.

Jutte, A. (2024, July). Explainable MLOps: A methodological framework for the development of Explainable AI in practice. In *The 2nd World Conference on eXplainable Artificial Intelligence*, Valletta, Malta.

Katreddy, Srinivasa Subramanyam. (2022). Robust MLOps frameworks for automating the AI/ML lifecycle in cloud environments. *International Journal of Intelligent Systems and Applications in Engineering*, 10(3s), 307–316. https://ijisae.org/index.php/IJISAE/article/view/7377

Kreuzberger, D., Kühl, N., and Hirschl, S. (2023). Machine learning operations (MLOps): Overview, definition, and architecture. *IEEE Access*, 11, 31866–31879.

Lakarasu, P. (2022). MLOps at scale: Bridging cloud infrastructure and AI lifecycle management. Available at SSRN 5272259.

Mucha, T. M., Ma, S., and Abhari, K. (2022, August). Beyond MLOps: The lifecycle of machine learning-based solutions. In *American Conference for Information Systems*, Article 1012.

Nogare, D., and Silveira, I. F. (2025). MLOps for machine learning model lifecycle automation – A systematic literature review. Authorea Preprints.

Stone, J., Patel, R., Ghiasi, F., Mittal, S., and Rahimi, S. (2025, May). Navigating MLOps: Insights into maturity, lifecycle, tools, and careers. In *IEEE Conference on Artificial Intelligence (CAI)*, Santa Clara, CA (pp. 643–650). IEEE.

Tatineni, S., and Rodwal, A. (2022). Leveraging AI for seamless integration of DevOps and MLOps: Techniques for automated testing, continuous delivery, and model governance. *Journal of Machine Learning in Pharmaceutical Research*, 2(2), 9–41.

6 XAI in Model Training

6.1 INTRODUCTION

6.1.1 Integration of XAI Methods in Model Training

In the DL models, due to their black-box nature, bias and overfitting that often happen during training will not be noticed. The integration of XAI in the training phase can expose these issues by tracking feature attribution trends and/or explanation drifts through epochs.

As explainability is a powerful mechanism for diagnosing why the DL models fail, behave inconsistently, or misclassify the data, software developers can understand and identify the model's architectural weaknesses, data leakage, or bias amplification by means of internal interpretation.

Hence, by integrating XAI methods during training pipeline at different stages such as bias monitoring, feature visualization, and debugging, one can easily detect biases, overfitting, model faults, etc.

6.1.2 Workflow of XAI Integration in Model Training

During training, bias and overfitting arise especially in higher-performing DL models. In such situations, XAI frameworks can expose them by tracking feature attribution trends or epoch-wise drift explanations. Figure 6.1 depicts the integrated workflow of XAI in the model training phase.

Integrating XAI in the training pipeline helps in the following analysis:

- Drift Analysis – By applying LIME or SHAP in the training phase, one can monitor the drift analysis at every few epochs, which will provide values to be used in validation dataset.
- Bias Detection – By comparing the explanations across the most important features in the dataset and less important features, it will be easier to find whether the less important features dominate the decision making.
- Regularization of Explanation – By incorporating a penalty term during the training phase, stable or fair feature attributions will be enforced. For instance, $L(\text{XAI}) = |\text{SHAP}t - \text{SHAP}t - 1| \wedge 2$.
- Identification of Overfitting – Monitoring with XAI methods during the training phase helps in periodical visualization of training outcome. This leads to focus on irrelevant features or areas in heatmaps, which will help in identifying overfitting.

DOI: 10.1201/9781003630241-8

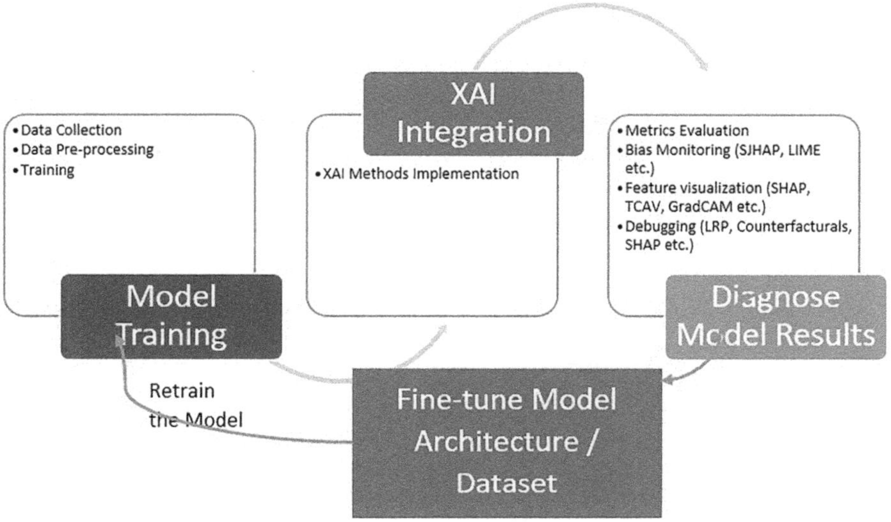

FIGURE 6.1 Workflow of XAI integration during model training phase.

6.2 FEATURE VISUALIZATION DURING TRAINING

To find which features have dominated the training phase, it is essential to get continuous visualization of features. This will help in fine-tuning the architecture of the model, pruning of features, and layer-wise weight adjustments before the training phase ends.

Some of the visualizations integrated in training phase are listed below:

- Feature Tracking – At every few epochs during the training phase, the computation of SHAP values for every feature and visualizing them as feature contribution trend graphs will help in identifying the features that are dominating the decision.
- Activation Mapping – The application of GradCAM in image processing applications in each layer of CNN, will help in visualizing CNN feature maps. By observing the convergence of the attention regions in semantically relevant areas, one can easily track the training phase. If there is any drifting or noise in the heatmap, it suggests to refine the architecture of the model.
- Visualizing Concept Activation Vector (TCAV) – By applying TCAV during the training phase, one can find how strongly each concept influences the final decision/outcome from the model. It helps to ensure that the model aligns with domain knowledge instead of latent artifacts.
- Interactive Dashboards – By providing interactive dashboards in the training phase, real-time visualization of XAI metrics can be done by means of TensorBoard with SHAP visualizer. This will help in monitoring interpretability along with the traditional model metrics such as accuracy, precision, etc.

6.3 XAI INTEGRATION STRATEGIES

During the learning phase, the XAI integration can be done using the following strategies:

- Learning Focus using GradCAM visualization – In deep neural network models such as CNNs, the attention maps based visualizations will map the key layers during the training phase. It helps to ensure the learning process progresses from generic features such as edges/textures to more specific features such as objects/ concepts.
- Dynamic Tracking using SHAP – In this strategy, for every few epochs, the SHAP value will be calculated for each feature. The visualization of trends will help in finding the important features that impact the decision. The features with increasing SHAP values in the trend graphs will be reinforced. The features with decreasing values may be analyzed for redundant features.
- Representation of Domain Relevancy using CBMs – In the case of image processing applications, this will help introduce concept layers in-between to represent domain-relevant features such as tumor density, eye vessel structure, etc. After introducing these layers, SHAP values are calculated to ensure these concepts impact the prediction explanations.

6.4 XAI TECHNIQUES FOR MODEL DEBUGGING

By means of interpreting the internal working of the models, the developers can easily find the weaknesses of the model architecture, data leakage, or bias amplification.

6.4.1 XAI Integration Approaches

Some of the integration approaches are:

- Counterfactual Explanations – By generating minimal changes (one change at a time) in the input features and finding their impact on the output class, one can debug decision boundaries to improve generalizations.
- Attribution Outlier Analysis – By identifying the samples that have huge changes from the dataset norm value using SHAP/LIME explanations, one can find the mislabeled, noisy, or adversarial samples from the dataset.
- Layer-wise and Neuron-wise Analysis – By applying layer-wise feature attribution techniques, one can find how each neuron in each layer contributes to the overall output. This helps in debugging the inactive neurons or incorrect gradient flow during backpropagation. Also, it helps in finding the neurons that contribute correctly by means of consistent explanations derived from the XAI methods.
- Guided Error Analysis – By comparing the correct and incorrect predictions using explanations, one can find which features are irrelevant and leading to wrong predictions. It helps in finding feature misalignment.
- Visual Debugging – Some of the guided XAI methods such as GradCAM can help to find CNN filters that latch onto dataset artifacts.

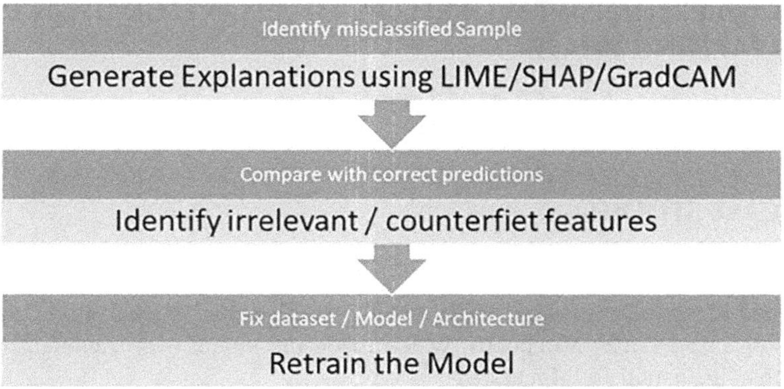

FIGURE 6.2 Debugging loop for DL models with XAI integration.

6.4.2 XAI Techniques to Debug DL Models

As DL models are black boxes in nature, it will be very difficult to find the reason for model failure. These models fail without any intimation in terms of misclassification of samples, specific region overfitting, or generating unstable predictions.

In these situations, XAI techniques can act as a debugging mechanism to find why the model has failed and where its reasoning went wrong.

The debugging loop of XAI integration is given in Figure 6.2.

6.5 FEATURE IMPORTANCE VISUALIZATION DURING TRAINING

The feature importance visualization during training phase helps in obtaining information on which features, neurons, or concepts are impacting the predictions at each epoch. It helps in early detection of weak or redundant features and neurons. Using this information, it will be easier to dynamically fine-tune layer weights and architecture.

It is given in the following workflow diagram (see Figure 6.3).

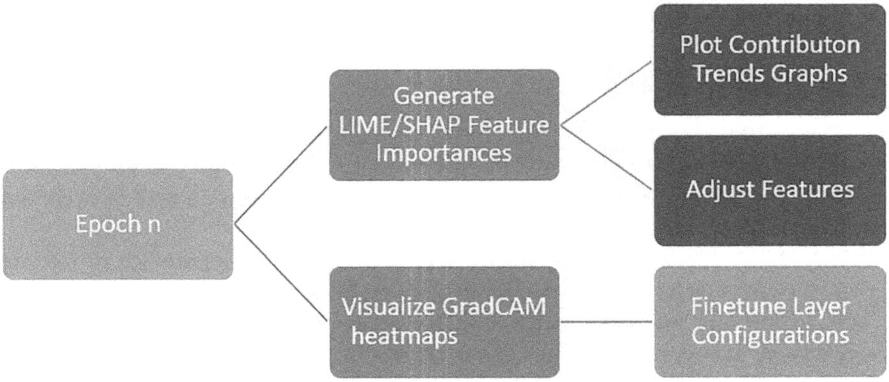

FIGURE 6.3 Workflow of feature visualization during training phase.

6.6 COMPREHENSIVE WORKFLOW OF XAI INTEGRATION IN LEARNING PHASE

The complete integration of XAI in learning phase is given in the comprehensive workflow diagram as shown in Figure 6.4.

6.7 CASE STUDIES

6.7.1 Case Study 1

In a facial recognition system, SHAP values revealed that "background color" had increasing impact after 50 epochs, which made the model overfit camera lighting rather than the actual facial features.

Using the SHAP values, this was revealed, and the remedial step is to perform data augmentation with illumination normalization and attention refocusing by the developers to resolve this issue.

6.7.2 Case Study 2

In a Diabetic Retinopathy Detection System, GradCAM maps are visualized for every 5 epochs. Using the visualization, it is identified that during the earlier epochs the model has focused on the image corners and labels. Based on this, the model is fine-tuned in its architecture level by adding more attention layers and is then localized on more specific features to be considered for such prediction.

6.7.3 Case Study 3

In a Pneumonia Detection system, the CNN model was identified with several misclassifications. The mitigation step is to apply GradCAM on these misclassified images and generate heatmaps. The heatmaps revealed the model's focus is on X-Ray frame edges and patient labels but not on lung opacity regions.

On analyzing this visualization, the preprocessing step was fine-tuned and the final convolutional layer was reinitialized to fix the bias. This improved both accuracy and explainability.

6.7.4 Case Study 4

In a Covid-19 detection model based on X-Ray images, the researchers have applied GradCAM and SHAP to monitor the training phase. Based on the visualizations derived, it is identified that the CNN model initially focused on text markers and scanning artifacts and not the lung opacity.

After 30th epoch, the SHAP values indicated increasing importance of relevant pixel clusters, which indicated that there was overfitting.

On identifying the flaws using XAI, the researchers retrained the model with cropped and standardized images. As a result, the later application of GradCAM heatmaps, indicated a change of focus on lung regions and not on the text markers or other artifacts. This improved the accuracy of the model and the overall model's performance.

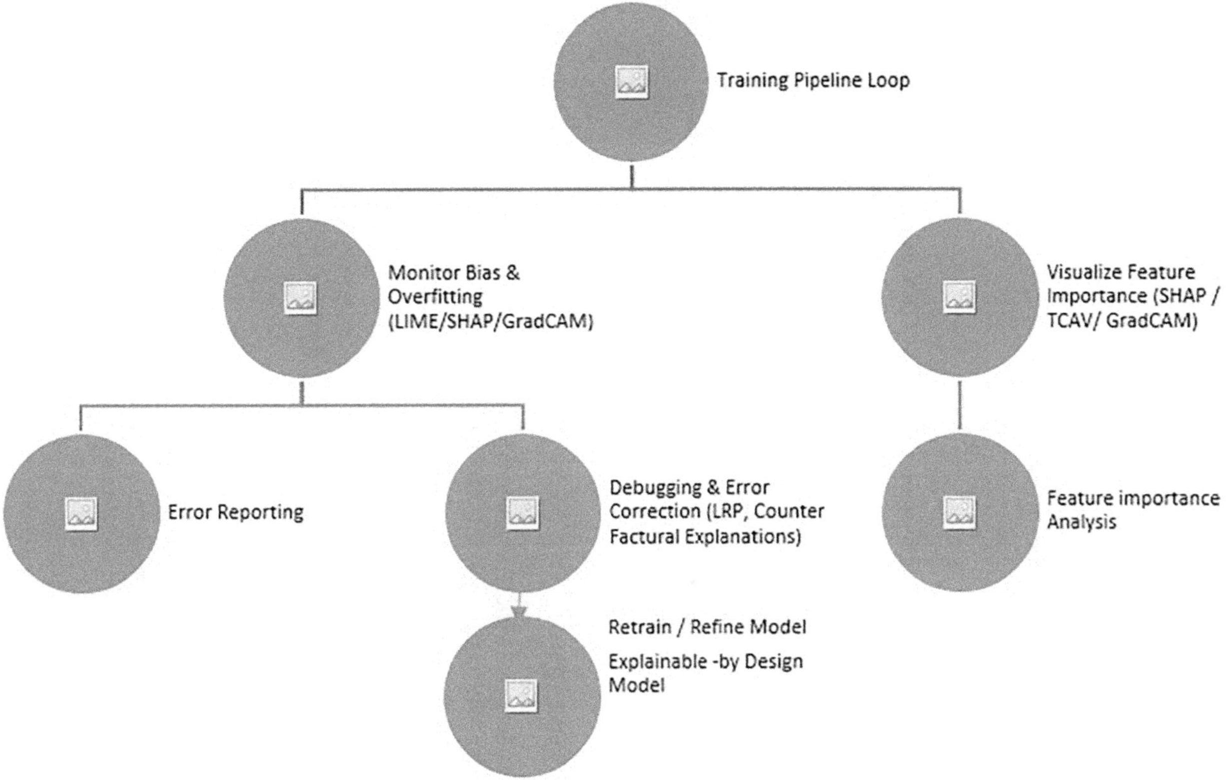

FIGURE 6.4 Comprehensive workflow of XAI integration in learning phase.

6.7.5 Case Study 5

In a Diabetic Retinopathy Severity Detection system, GradCAM is applied during training and heatmaps are visualized for every 10 epochs. The heatmaps revealed that, initially, the CNN model had taken the retina border.

As training advanced by adding more attention modules and training for further epochs, the heatmap visualizations indicated a shift of focus on microaneurysm regions. In addition, the SHAP values showed increased importance for "lesion texture features." Continuous visualizations using XAI methods will provide a valid and reliable model with interpretability and clinical relevance.

6.7.6 Case Study 6

In the Pneumonia Detection system, the applied CNN model misclassified healthy X-Rays as Pneumonia. Then, the GradCAM method is applied on the misclassified images. The generated heatmaps showed the activation is on the image borders and text labels and not on lung tissues.

The analysis indicated that the text annotations are correlated with pneumonia samples, which led to a data leakage bias.

On getting this analysis result, the images are cropped and augmentations is applied to remove such artifacts. Then the model is retrained and further application of GradCAM generated heatmaps indicated the realignment of lung opacity regions which confirms correct prediction by the model.

6.8 SUMMARY

As DL models are black boxes in nature, bias and overfitting that often happen during training will not be noticed. The integration of XAI in the training phase can expose these issues by tracking feature attribution trends and/or explanation drifts through epochs.

As explainability is a powerful mechanism for diagnosing why the DL models fail, behave inconsistently, or misclassify the data, software developers can understand and identify the model's architectural weaknesses, data leakage, or bias amplification by means of internal interpretation.

BIBLIOGRAPHY

Ahmad, M. A., Eckert, C., and Teredesai, A. (2018). Interpretable machine learning in healthcare. In *Proceedings of the 2018 ACM International Conference on Bioinformatics, Computational Biology, and Health Informatics*, pp. 559–560. https://doi.org/10.1145/3233547.3233667

Apicella, A., Giugliano, S., Isgrò, F., and Prevete, R. (2025). IMPACTX: Improving model performance by predicting correct explanations. arXiv.

Baniecki, H., and Biecek, P. (2023). Adversarial attacks and defenses in explainable artificial intelligence: A survey. arXiv preprint arXiv:2306.06123.

Bhati, M., Jain, R., and Raghuwanshi, B. S. (2024). Explainable artificial intelligence for medical image analysis: A survey. *Journal of Imaging*, 10(10), 239. https://doi.org/10.3390/jimaging10100239

Das, A., and Rad, P. (2020). Opportunities and challenges in explainable artificial intelligence (XAI): A survey. arXiv preprint arXiv:2006.11371.

Doshi-Velez, F., and Kim, B. (2017). Towards a rigorous science of interpretable machine learning. arXiv preprint arXiv:1702.08608.

Hou, J., Liu, S., Bie, Y., Wang, H., Tan, A., Luo, L., and Chen, H. (2024). Self-eXplainable AI for medical image analysis: A survey and new outlooks. arXiv preprint arXiv:2410.02331.

Khadivpour, F., Banerjee, A., and Guzdial, M. (2022). Responsibility: An example-based Explainable AI approach via training process inspection. arXiv preprint arXiv:2209.03433.

Kim, B., Wattenberg, M., Gilmer, J., Cai, C., Wexler, J., Viégas, F., and Sayres, R. (2018). Interpretability beyond feature attribution: Quantitative testing with concept activation vectors (TCAV). In *Proceedings of the 35th International Conference on Machine Learning*, Stockholm, Sweden, pp. 2673–2682.

Koh, P. W., and Liang, P. (2017). Understanding black-box predictions via influence functions. In *Proceedings of the 34th International Conference on Machine Learning*, Sydney, Australia, pp. 1885–1894.

Linardatos, P., Papastefanopoulos, V., and Kotsiantis, S. (2021). Explainable AI: A review of machine learning interpretability methods. *Entropy*, 23(1), 18. https://doi.org/10.3390/e23010018

Lundberg, S. M., and Lee, S.-I. (2017). A unified approach to interpreting model predictions. *Advances in Neural Information Processing Systems*, 30, 4765–4774.

Nguyen, T. T. H. (2025). A survey on XAI-based model improvement methods (Draft PDF).

Olah, C., Mordvintsev, A., and Schubert, L. (2017). Feature visualization. *Distill*. https://doi.org/10.23915/distill.00007

Ribeiro, M. T., Singh, S., and Guestrin, C. (2016). "Why should I trust you?" Explaining the predictions of any classifier. In *Proceedings of the 22nd ACM SIGKDD International Conference on Knowledge Discovery and Data Mining*, pp. 1135–1144. https://doi.org/10.1145/2939672.2939778

Samek, W., Wiegand, T., and Müller, K.-R. (2017). Explainable artificial intelligence: Understanding, visualizing and interpreting deep learning models. arXiv preprint arXiv:1708.08296.

Selvaraju, R. R., Cogswell, M., Das, A., Vedantam, R., Parikh, D., and Batra, D. (2017). Grad-CAM: Visual explanations from deep networks via gradient-based localization. In *Proceedings of the IEEE International Conference on Computer Vision*, pp. 618–626. https://doi.org/10.1109/ICCV.2017.74

Simonyan, K., Vedaldi, A., and Zisserman, A. (2014). Deep inside convolutional networks: Visualising image classification models and saliency maps. arXiv preprint arXiv:1312.6034.

Springenberg, J. T., Dosovitskiy, A., Brox, T., and Riedmiller, M. (2015). Striving for simplicity: The all convolutional net. In *International Conference on Learning Representations (ICLR)*, California.

Zheng, Q., Wang, Z., Zhou, J., and Lu, J. (2022). Shap-CAM: Visual explanations for convolutional neural networks based on Shapley value. arXiv preprint arXiv:2208.03608.

7 XAI in Model Validation

7.1 INTRODUCTION

The evaluation of DL models is essential in order to deploy them in the target systems. XAI plays a vital role in providing insights on the working of such DL models by providing explanations and computations from the models' outcomes. However, the XAI methods' explanations need to be evaluated for their acceptance in the model's understanding. Hence, this chapter deals with the evaluation of explanations based on fidelity metrics.

The XAI methods such as SHAP, LIME, GradCAM, etc. are evaluated based on the various metrics such as fidelity, unambiguity, and interpretability size. This chapter takes a benchmark dataset and applied some of the DL models and then applied XAI methods on them. Finally, the explanations' evaluations are computed for each of the models to get the insight on the explanations applicability.

Due to the increasing complexity of DL models in real-time complex systems such as healthcare, finance, and automation, the reliability of explanations is highly important in order to accept the DL models based on their explainability. The infidelity measures are crucial for identifying discrepancies between model explanations and the underlying decision-making mechanisms, which can highlight potential areas of improvement or reveal limitations in the explanation methods.

7.2 EXPLAINABILITY METRICS FOR MODEL EVALUATION

As mentioned, the traditional performance metrics such as accuracy, precision, recall, and F1-Score don't have the ability to assess the trustworthiness of the model. Especially, they do not provide an understandable internal processing applied by the model to provide the prediction.

Some of the explainability metrics applied to evaluate the model are: interpretability, feature relevance consistency, and explanation stability. Their objectives and detailed descriptions are provided in Table 7.1.

EXAMPLE:

For an "AI-based Loan Approval System," SHAP values are analyzed to ensure that "Income amount" and "loan history" should contribute more than "Pincode." If the faithfulness value is lower and instability is higher, it indicates unreliable explanations.

 DOI: 10.1201/9781003630241-9

TABLE 7.1
Explainability Metrics

Name of the Metric	Objective	Application
Fidelity	To measure how closely the surrogate explanation model such as LIME or SHAP mimics the black-box model.	If the SHAP explanation differs greatly from the model predictions, then fidelity metric will be low. It indicates that the model is not trustworthy.
Stability	To measure the impact on the model's outcome when a small change/perturbation is made in the feature values.	In image processing applications, a small perturbation in the image may have impact on the model's prediction. This metric helps to evaluate the robustness of such models under a perturbed input.
Sparsity	To evaluate how concise and understandable the explanation is.	A typical example is LIME's number of features used in local approximation vs. the actual features used by the model for its prediction.
Comprehensibility Score	To evaluate the level of understanding by humans from the explanations produced.	In Medical AI, it is always essential to evaluate the trustworthiness of the model by experts in that field by getting the rating based on explanations.
Faithfulness Correlation	To find the correlation between feature importance and model output.	In SHAP validation, this correlation metric will be applied to find the faithfulness.

7.3 BALANCING ACCURACY AND INTERPRETABILITY

The black-box models such as deep neural network models generally exhibit high performance but they don't exhibit their internal working. Hence, XAI frameworks help to embed interpretability constraints or interpretable surrogates for the black-box models.

The following types of evaluations help to balance the accuracy and interpretability of a model:

1. Hybrid Evaluation – It helps to evaluate both the model's performance metrics such as accuracy, precision, recall, etc. along with XAI metrics such as fidelity, sparsity, etc.
2. Simplify Black-Box Model – It helps to convert a deep neural network-based model into a simpler decision tree while maintaining fidelity.
3. Human-in-the-Loop Validation – In healthcare analytics, the domain experts (doctors who are experts in that disease) will validate the explanations.

4. Multi-objective Optimization – Both prediction loss and interpretability loss are optimized so that we can get the relevant insight as below:

$$L = L\_\text{pred} + \lambda^{*} L\_\text{explain}$$

where

$L_$pred is the actual model outcome
$L_$explain is the explainability method outcome.

Sample:
In "Brain Tumor Analysis and Prediction," a deep neural network model such as VGG-16 achieves 98% accuracy, but GradCAM reveals the model has focused mainly on image borders and not on the exact features to be used for prediction. Using this as the feedback, attention regularization is added and based on that the accuracy is reduced to 96% but the interpretability and trustworthiness evaluations have increased significantly.

7.4 ALGORITHM FOR XAI EVALUATION IN DL MODELS

Steps involved in the model's validation are given below:

Step 1. Apply the DNN models for image classification.
Step 2. Analyze the traditional performance metrics such as accuracy, precision, recall, and F1-Score for all these models and find the best model with higher metric values.
Step 3. Analyze XAI metrics as below:
- Step 3.1. Generate explanations for the best model chosen using XAI methods such as SHAP and GradCAM on misclassified samples.
- Step 3.2. Identify the cause of failure by the following evaluations:
  - If the SHAP shows strong weight value for "less weighted feature" instead of "higher important feature," then the model has to be retrained.
  - If GradCAM highlights "less weighted image part" instead of "contributing part of the image," then it shows the model is failed in this area. And hence it needs to be retrained on this feature.
- Step 3.3. Diagnose the Model Bias:
  - If the model shows bias, it indicates data imbalance – hence it is time to analyze the dataset and do the training again.
- Step 3.4. Implement Fixes based on the above steps:
  - Make the dataset balanced by adding more samples.
  - Retrain and validate the model with bias monitoring.
  - Track the consistency of SHAP importance.

7.5 CASE STUDY

In this study, we aim to detect pneumonia using Deep Learning models and enhance the interpretability of the models through various explanation techniques. We employ three pretrained models – DenseNet, ResNet, and VGG – fine-tuning them to adapt to the pneumonia dataset (see Figure 7.1).

To understand and interpret the models' decision-making processes, we integrate explanation techniques such as SHAP (SHapley Additive exPlanations), LIME (Local Interpretable Model-Agnostic Explanations), and GradCAM (Gradient-weighted Class Activation Mapping).

7.5.1 Data Preprocessing

- Input Data: Chest X-ray images for pneumonia detection.
- Preprocessing Stage: The original images are of varying size. Hence, all the images are resized as 224 × 224. Then they are normalized and augmentations are applied to enhance the images for proper training.

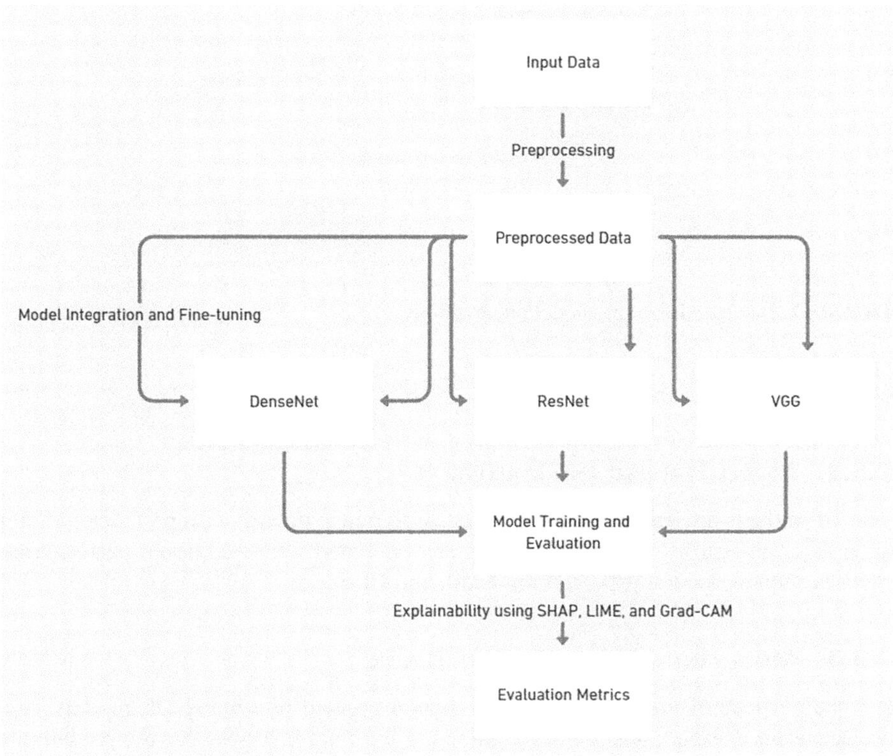

FIGURE 7.1 Process flow of integrated XAI evaluation of DL models.

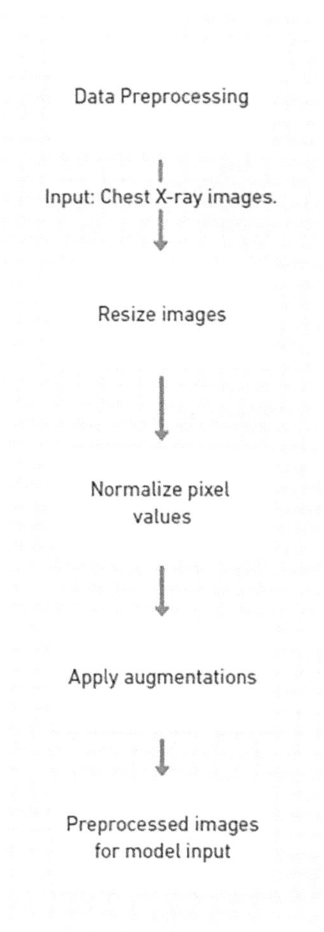

FIGURE 7.2 Deep Learning model process flow.

The process flow of the Deep Learning model is given in Figure 7.2.

7.5.2 Model Training and Evaluation

The DL models are trained on the preprocessed image dataset with parameter values such as cross-entropy loss. The models are evaluated with traditional performance metrics such as accuracy, precision, recall, and F1-Score.

7.5.3 Model Integration and Fine-Tuning

For the purpose of understanding, this chapter applied pretrained DL models such as DenseNet, ResNet, and VGG by adjusting the final layers as per the pneumonia dataset. It is shown in Figure 7.3.

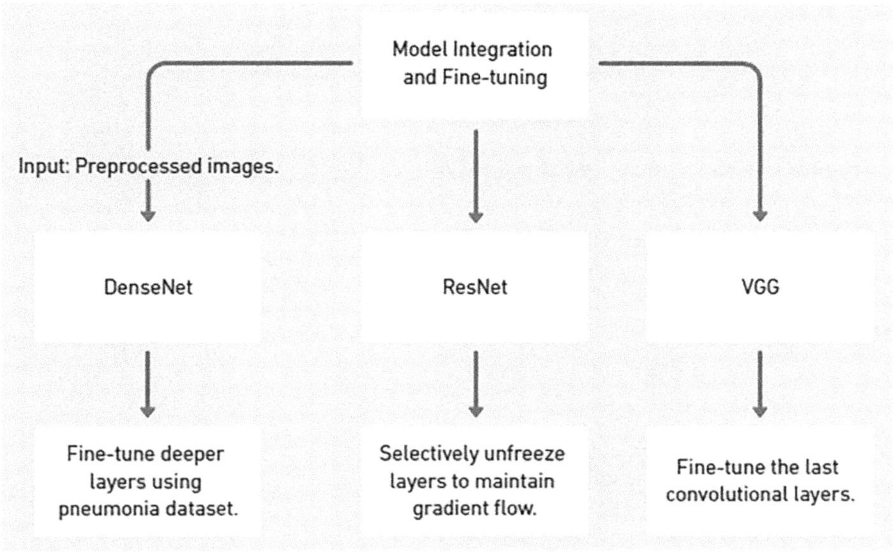

FIGURE 7.3 Workflow of applying various Deep Learning models.

7.5.4 XAI-Based Model Explanation

The following XAI methods are applied to get explanations on the model's internal working mechanism.

- LIME – This XAI method changes the input data, which is called perturbation, and applies the DL model and the XAI method to generate explanations. This helps in understanding the local decision behavior of the DL model by visualizing the most critical image regions.
- SHAP – This XAI method helps in computing the contribution of each feature in the final decision and also generates heatmaps that highlight the important regions for prediction.
- GradCAM – It produces the activation maps from the input images to localize the important regions that determines the decision-making.

7.5.5 XAI Evaluation Metrics

The following metrics are applied to evaluate the XAI methods' outcomes (see Figure 7.4).

- Infidelity – It measures the trustworthiness of the explanations in reflecting the actual behavior of the DL model.
- RPD (Representer Point Decomposition) – It helps in identifying the features that are influencing the decision-making.
- Robustness – It helps in evaluating the stability of explanations under perturbations.

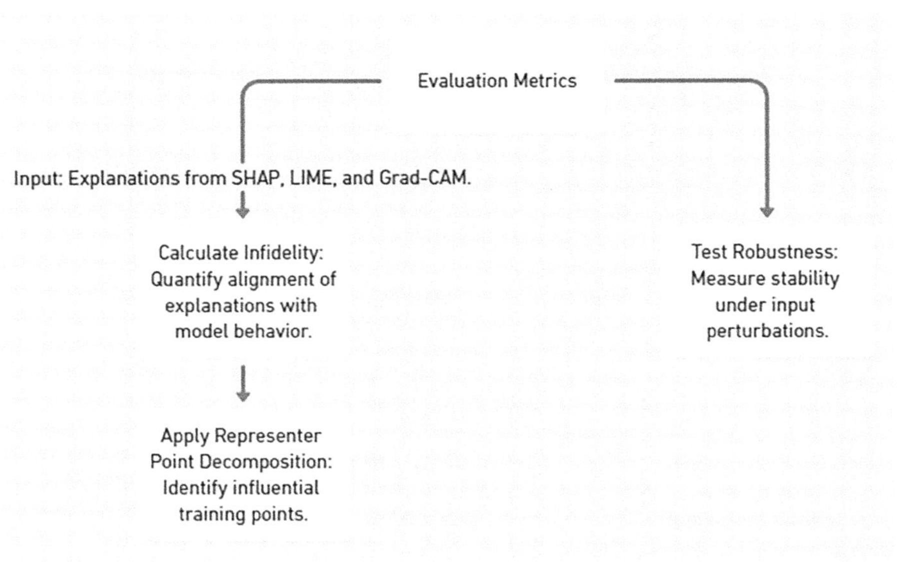

FIGURE 7.4 Metrics evaluation workflow for various XAI methods.

7.6 IMPLEMENTATION

7.6.1 Deep Learning Models Implementation

The pretrained DL models such as DenseNet, ResNet, and VGG are implemented and the metrics are evaluated for their performance assessment. The results are given in Table 7.2.

7.6.2 XAI Implementation

The XAI methods LIME, SHAP and GradCAM are implemented on the DL models and their outcomes were evaluated using metrics such as infidelity, robustness and RPD metrics.

TABLE 7.2
Model Accuracies for Pneumonia Detection on Test Data

S.No.	Model Used	Accuracy
1	VGG	78%
2	ResNet	82.53%
3	DenseNet	85.73%

7.7 EVALUATION OF DL MODELS USING XAI VISUALIZATIONS

The evaluations of XAI methods are tabulated in Table 7.3. Apart from that, the XAI outcomes such as heatmaps and activation maps are generated to show the key regions of the lung images during pneumonia prediction. Then, a comparative analysis is done to find which XAI method provides reliable explanations.

LIME provides a holistic view of feature contributions. While LIME offers localized interpretations. GradCAM effectively visualizes spatial regions for image-based models. SHAP is expected to be the most reliable in reflecting model behavior across all three architectures.

Infidelity in the context of evaluating Machine Learning model explanations refers to a metric that measures how well an explanation aligns with the actual behavior of the model when inputs are perturbed. The results indicated that VGG architecture is consistent for the explanations.

The results indicated that the LIME explanation from DenseNet and VGG are not providing the focus on most relevant regions. It is shown in Figure 7.5. The GradCAM based visualization is given in Figure 7.6.

TABLE 7.3
XAI Methods Evaluation

Metrics	VGG	DenseNet	ResNet
Infidelity using LIME	0.007541527505964041	0.0762767344713211	0.009043559432029724
Robustness using LIME	0.562767344713211	0.647132115627673	0.447132115627673

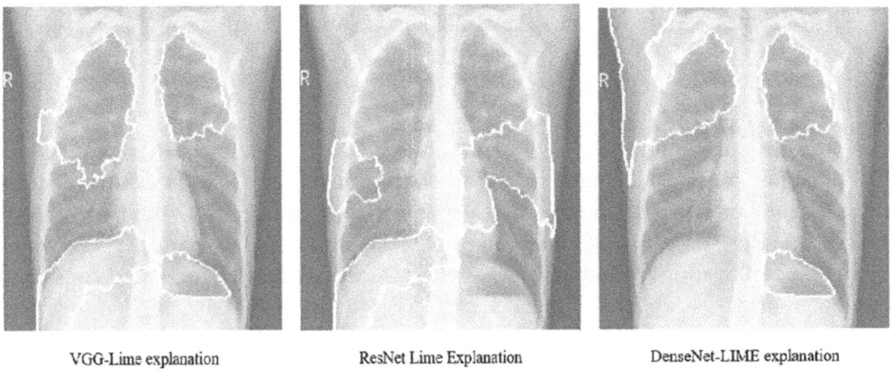

FIGURE 7.5 LIME visualization of key regions driving pneumonia prediction.

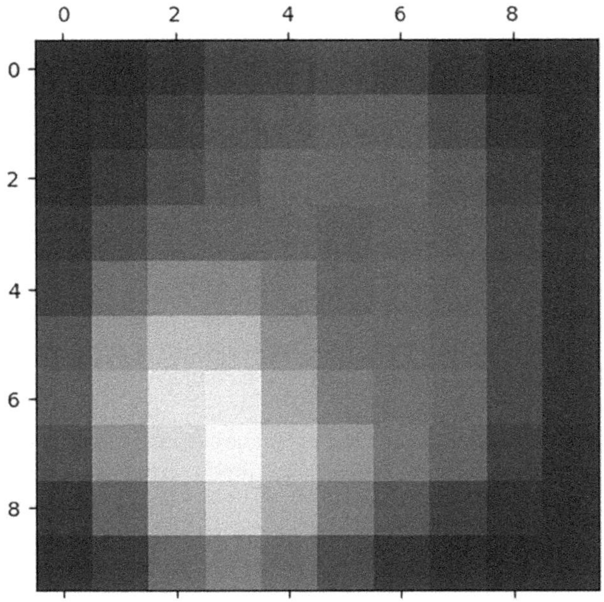

FIGURE 7.6 Visualization of key regions driving pneumonia prediction using GradCAM.

7.8 EVALUATION OF INTEGRATED XAI METHODS

The XAI methods (SHAP, LIME, and GradCAM) are evaluated based on their ability to explain predictions consistently and accurately. The evaluation considers:

- Explanation Fidelity: How well each method aligns with the actual model behavior.
- Interpretability: Ease of understanding the explanations provided by each method.
- Robustness: Stability of explanations under input perturbations.

7.9 SUMMARY

This chapter provides a comprehensive insight on the application and evaluation of XAI methods in DL models assessment. The pretrained DL models such as DenseNet, ResNet and VGG and found DenseNet has higher accuracy. However, when the explanations based analysis is done, the VGG model generated consistent explanations across epochs. The LIME, SHAP and GradCAM are applied to generate the explanations for all the models.

However, it is identified that each of the XAI methods have some strengths in each of these DL models. GradCAM showcases the spatial regions, LIME offers a detailed explanation on features contributions. The SHAP method gave the insights on the model's behavior. SHAP has demonstrated.

BIBLIOGRAPHY

Arrieta, A. B., Díaz-Rodríguez, N., Gómez, A., Hernández-Lobato, J. M., and Matía, J. (2020). A survey of explainable artificial intelligence (XAI): Toward more explainable models. *IEEE Access*, 8, 134405–134422.

Choi, S., and Lee, J. (2021). *Interpretable Machine Learning: A Guide for Making Black Box Models Explainable*. Springer.

Erion, G., Janizek, J. D., Sturmfels, P., Lundberg, S., and Lee, S. I. (2019). Learning explainable models using attribution priors. arXiv preprint arXiv:1906.10670.

Gilpin, L. H., Ribeiro, M. T., and Caruana, R. (2018). Explainable AI: A review of techniques and applications. *Proceedings of the AAAI/ACM Conference on AI, Ethics, and Society*, Louisiana. https://doi.org/10.1145/3278721.3278725

Hooker, S., Erhan, D., Kindermans, P. J., and Kim, B. (2018). Evaluating feature importance estimates. arXiv preprint arXiv:1806.10758.

Hsieh, C. Y., Yeh, C. K., Liu, X., Ravikumar, P., Kim, S., Kumar, S., and Hsieh, C. J. (2021). Evaluations and methods for explanation through robustness analysis. arXiv preprint arXiv:2006.00442.

Lipton, Z. C. (2018). The mythos of model interpretability. *Communications of the ACM*, 61(3), 36–43.

Miller, T. (2019). Explanation in Artificial Intelligence: Insights from the social sciences. *ACM Computing Surveys (CSUR)*, 51(3), 1–42. https://doi.org/10.1145/3285029

Petsiuk, V., Das, A., and Saenko, K. (2018). RISE: Randomized input sampling for explanation of black-box models. arXiv preprint arXiv:1806.07421.

Ribeiro, M. T., Singh, S., and Guestrin, C. (2016). Why should I trust you? Explaining the predictions of any classifier. *Proceedings of the 22nd ACM SIGKDD International Conference on Knowledge Discovery and Data Mining* (pp. 1135–1144).

Smith, J., and Johnson, L. (2022). Advancements in objective evaluation metrics for machine learning model explanations. *Journal of Artificial Intelligence Research*, 74, 15–34.

Tjoa, E., and Guan, C. (2020). A survey on explainable artificial intelligence (XAI): Toward medical XAI. *IEEE Transactions on Neural Networks and Learning Systems*, 32(11), 4793–4813.

Wachter, S., Mittelstadt, B. D., and Russell, C. (2017). Counterfactual explanations without opening the black box: Automated decisions and the GDPR. arXiv preprint arXiv: 1711.00399.

Yeh, C. K., Hsieh, C. Y., Suggala, A. S., Inouye, D. I., and Ravikumar, P. (2019). On the (in)fidelity and sensitivity of explanations. *Advances in Neural Information Processing Systems*, 32.

Yeh, C. K., Kim, J., Yen, I. E. H., and Ravikumar, P. K. (2018). Representer point selection for explaining deep neural networks. *Advances in Neural Information Processing Systems*, 31.

Yeh, C. K., and Ravikumar, P. (2021). Objective criteria for explanations of machine learning models. *Applied AI Letters*, 2(4), e57. https://doi.org/10.1002/ail2.57

8 XAI Integration in CI/CD

8.1 INTRODUCTION

As AI models are transitioning from research set-ups to large-scale production systems for industry-strength software development, the need for transparency, accountability, and trustworthiness becomes paramount.

The traditional CI/CD pipelines focus on automated code testing, quality assurance automation, as well as automation of containerization and deployment. However, the deployment of ML and DL models using these traditional CI/CD pipelines will never take into consideration the model's trustworthiness. Rather, they simply focus on functional correctness and evaluation of model metrics such as accuracy, precision, recall, etc.

Hence, it is time to evolve the CI/CD pipeline to integrate XAI to ensure every model pushed to production is not only accurate but also trustworthy, understandable, and compliant with ethical and regulatory standards.

Today's ML/DL workflow involves iterative improvement to data, features, hyperparameters, training pipelines, and model architectures. However, it must be noted that each iteration significantly impacts a model in making decisions. It is to be noted that without explainability evaluations, organizations will take the risk of deploying models based on the accuracy metric. However, as these models are black boxes, the results derived for one instance need not give the same trustworthy outcome for another instance as it may be impacted by some other features.

Hence, it is essential to embed XAI methods within CI/CD lifecycle in order to identify the model's vulnerability to biases, data drift, adversarial manipulation, and regulatory violations. This will provide a continuous and automated quality checkpoint for the model before deployment.

In addition, it addresses the growing global AI governance requirements such as the EU AI Act, ISO 42001, and industry guidelines in domains such as healthcare, finance, and autonomous systems.

8.2 ADVANTAGES OF CI/CD WITH XAI

It allows the developers to systematically validate the following:

1. Interpretation of model decisions by the application of XAI methods such as SHAP, LIME, GradCAM, Saliency Maps, etc.
2. Assessment of explainability metrics to check for explanation consistency.
3. Validation of model bias and fairness assessments at quality checking gates.
4. Provision of detailed explanations from the model to all stakeholders at different levels.

Thus, this transforms a one-time analysis into a continuous assurance mechanism as XAI is integrated into the CI/CD pipeline.

 DOI: 10.1201/9781003630241-10

8.3 CI/CD INTEGRATION WITH XAI

In CI/CD, the CI (Continuous Integration) performs automated explainability tests before accepting any new model iteration while the CD (Continuous Deployment) pipeline ensures real-time explainability for models that are seen by the end users. Hence, it develops a closed-loop system where transparency is maintained at every level of the development lifecycle.

8.3.1 Components of XAI-Enabled CI/CD Ecosystem

The XAI-integrated CI/CD ecosystem should contain the following components:

- Automated evaluation of explainability during model validation.
- Storing of version-controlled explanation artifacts along with model executables.
- Deployment of real-time interpretability services as microservices or edge components.
- Dashboard monitoring to track explanation drift, adversaries, attribution changes, and anomaly patterns.
- Ensuring ethical and governance frameworks for auditability and reproducibility.

8.3.2 Process Flow of XAI Integration in CI/CD

The process flow diagram of XAI integration in CI/CD pipeline is given in Figure 8.1. The major two parts, Continuous Integration (CI) and Continuous Delivery (CD), are

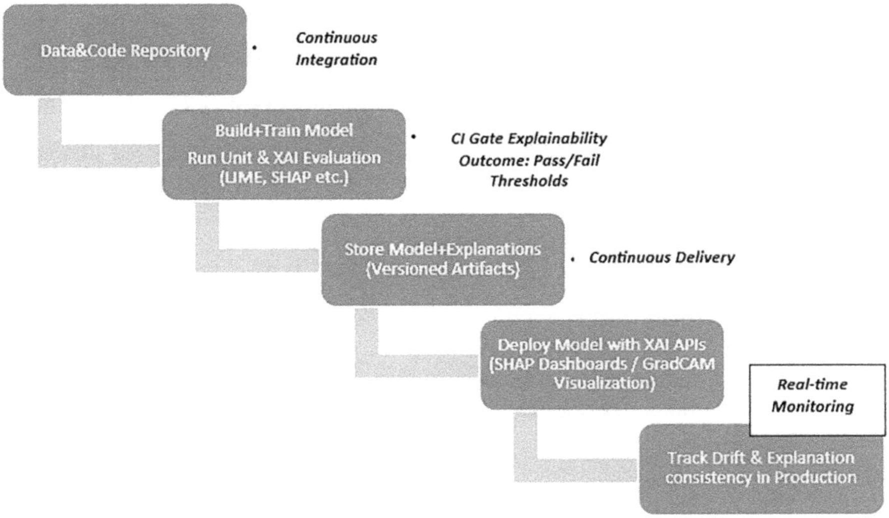

FIGURE 8.1 Process flow diagram of XAI-integrated CI/CD pipeline.

TABLE 8.1
XAI tools for CI/CD process flow

Phase/Testing	Objective	Tools
Continuous Integration/Automated Interpretability Testing	Ensures every model built is explainable	SHAP, LIME, PyTest, MLflow
CI Gate for Explainability	Block nontransparent Models	Custom CI Scripts, Jenkins
Continuous Delivery/Explainability Delivery	Deploy models with explanation lineage	DVC, MLflow, SageMaker Clarify
Real-time Interpretability Testing	Monitor live explainability drift	Fiddler AI, Arize AI, Evidently

now provided with XAI components. The CI component has 'Build + Train' model which runs both unit testing for 'Build' and XAI-based evaluation using XAI methods such as LIME, SHAP, etc. to get explainability outcomes apart from regular pass/fail test results. To achieve it, a CI gate explainability component is integrated in the CI pipeline.

In the case of CD, while storing the versioned artifacts, the explanations are also stored along with the DL model implementation. In addition to that, the XAI methods are provided in the XAI dashboards to choose which kind of explanation is needed from the model such as textual or graphical explanations. During the time of production, a real-time monitoring utility is added to track drifts and explanation consistency before deployment.

The objective and the tools to be used for the above process flow are given in Table 8.1.

Let us discuss each of these phases in detail in the subsequent sections.

8.4 EXPLAINABILITY IN CONTINUOUS INTEGRATION (CI)

8.4.1 Automated Interpretability Testing During Model Development

During CI, automated pipelines are built to test models continuously. At this stage, XAI integration allows automated interpretability evaluations to ensure every model build maintains and improves the explainability of the model's predictions.

Implementation Stages:

1. Integrate explainability libraries such as SHAP, LIME, etc. into CI test scripts.
2. Compare baseline explanation distributions with new model explanations.
3. Define thresholds for fidelity, stability, and feature-importance drift.

Sample:
Consider an "AI-based Fraudulent Transaction Monitoring System." The ML/DL model is retrained weekly to get updated training on new threats and transactions

made. In this model, a SHAP-based test will help in finding whether the model is appropriately trained. It is done as below:

8.4.1.1 Phase 1: Unit Testing for Transparency

- The CI phase computes SHAP importance for top features.
- If the new model's explanations deviate greater than 10% from baseline, for instance – the "transaction amount" feature loses its importance and "account number" gains higher weightage – then the build fails the interpretability test.
- This indicates that the model needs to be retrained and its validation needs to be assessed for biases, adversarial attacks, etc.

8.4.1.2 Phase 2: Explainability as a CI Gate for Model Progression

In MLOps, the "CI gates" act as evaluation checks of the ML/DL model which must pass before moving to the next step. XAI metrics can serve as one such gate to block opaque or biased models.

There are three types of CI gates namely:

- Performance Gate – It will check the performance metrics such as accuracy, precision, recall, etc. (Example: accuracy > 95%)
- Bias Gate – It will check the bias of the model using XAI methods (Example: SHAP parity gap <= 5% between subgroups)
- Explainability Gate – It will check the explainability metrics (Example: Fidelity >= 0.9, Stability >= 0.85, etc.)

These gates are used to perform the appropriate checks related to the model.

Case Study:

Consider the same "AI-integrated Fraudulent Transaction Monitoring System."
The CI gate helps to ensure that the feature attributions remain consistent. For instance, "transaction amount" should always be explained with > 40% of predictions.
If LIME or SHAP shows a drop or shift in key features in these explanations, CI halts the pipeline and flags the build.

8.4.2 Explainability in Continuous Delivery (CD)

8.4.2.1 Versioning of Explanations

Generally, in CI/CD, version control is applied to code and other artifacts in the repository. In the case of AI-based systems, not only the ML/DL model code but also the model weights, configurations, and explanations must be versioned.

Some of the explainability artifacts including SHAP summary reports, GradCAM visualizations, LIME graphs, etc. should be versioned to support usability and regulatory review.

To ensure this, some of the best practices to be used are:

1. Store explanation metadata such as feature importance, heatmaps, etc. apart from model executable files in version control repositories such as MLflow, DVC, etc.
2. Deliver explanations as part of the model package to APIs and dashboards.
3. Map each model version to its corresponding interpretability versions with metrics to maintain explanation lineage.

Case Study:

Consider an "AI-integrated Brain Tumor Prediction System." For this model, each version of the model pushed for checking includes:

- GradCAM visualizations for validation images and LIME summary reports.
- Model registry entries with metrics such as fidelity, interpretability size, unambiguity, etc.

Keeping all these in the repository helps to ensure reproducibility and regulatory traceability such as FDA and EU AI Act compliance.

8.4.2.2 Real-Time Interpretability Testing in Production Environment

Once the deployment is done, the deployed models must maintain explainability under real-time environmental data. Hence, real-time interpretability testing evaluates whether the explanations associated with the model are stable and understandable when there is a data drift or concept drift.

Some of the techniques to be applied are:

- Online SHAP monitoring – Here, the SHAP explanations are computed on live predictions to detect feature drift.
- Explainability monitor dashboards – These will track trends in explanation fidelity and fairness.
- Automated alerts on deviations – There should be automated triggers to give alerts when explanations deviate from expected patterns.

Case Study:

Consider an "E-Commerce Recommender System." After deployment in the real-time environment, if the LIME explanations showcase the feature "price" as the higher-impact feature on the outcome, while the feature "user preference," which was the original higher-impact feature before deployment, indicates interpretability drift, it will be triggered automatically for human inspection to ensure the model's fidelity.

8.5 XAI INTEGRATION WITH CI/CD – TOOLS AND FRAMEWORKS

This section gives a sample set of tools and frameworks that are applicable for XAI integration in CI/CD pipeline as given in table 8.2.

TABLE 8.2
XAI tools and their purpose in CI/CD integration

Tools/Frameworks	Purpose	XAI Integration in CI/CD
MLflow/Kubeflow	Used for model versioning and pipeline orchestration	They help in storing and comparing SHAP and LIME reports as artifacts.
Data Version Control (DVC)	Used to version the data, models, and metrics	Helps in tracking metadata and feature drift.
Evidently AI	Used to monitor model and detect drifts	Provides explainability dashboards and feature impact tracking.
TensorBoard	Applies training and validation visualization	Integrates GradCAM, SHAP, and summary plots to get a complete insight.
Seldon Deploy/Fiddler AI/ Arize AI	Provides MLOps + Explainability as service	Generates real-time explanations and bias reports.
Amazon SageMaker Clarify/ Azure InterpretML	Manages all XAI frameworks	Automates the SHAP, bias, and feature attribution in deployment.

8.6 SAMPLE TOOL FOR XAI INTEGRATION IN CI/CD

The tool taken for the case study is Amazon SageMaker Clarify. The case study is on "Loan Eligibility Prediction." The case study is to apply an ML model in a banking application to predict whether a person's loan application can be approved or not. The CI/CD pipeline for this case study must ensure:

1. No age or gender bias is introduced during training.
2. Across versions, the feature importance must be stable.
3. For the end users – in the case study, loan sanction officials – proper explanations should be provided.

8.6.1 XAI Tool Integration in CI/CD Pipeline

Let us now discuss the steps involved in applying "Amazon SageMaker Clarify" – an XAI tool – to perform the CI/CD flow with XAI.

8.6.1.1 Step 1 – Continuous Integration (CI) – Model Training

Here, the model training is done through CodePipeline/GitHub Actions/Jenkins. Once the training is completed, the following code has to be provided.

```
from sagemaker import clarify
clarify_processor = clarify.SageMakerClarifyProcesso
    r(role=role, instance_count=1, instance_type='ml.
    m5.xlarge', sagemaker_session=session)
```

8.6.1.2 Step 2 – Pretraining Bias Analysis

Now, to make Clarify analyze the dataset bias, the following code has to be fed:

```
bias_config = clarify.BiasConfig(label='loan_approved',
   facet='gender', threshold=0.2)
```

Based on this, the Clarify tool will generate metrics such as:

1. Disparate Impact Ratio
2. Difference in Positive Prediction Rate

The result will be stored in S3.

8.6.1.3 Step 3 – Post-Training Bias with SHAP Explainability

In this step, the Clarify tool evaluates the trained model using the following code:

```
explainability_config = clarify.
   SHAPConfig(baseline=[0,0,0], num_samples=100)
```

Once it is set up, the Clarify tool generates the following:

1. Global SHAP bar charts to display the highest-impacting features
2. Local explanations for each instance, that is, per loan applicant
3. Explanations in the form of JSON files for dashboards

8.6.1.4 Step 4 – Pipeline Quality Gate

In this step, the CI/CD pipeline validations are going to be carried out. During CI/CD pipeline, the following validations are performed and based on that, the decision on whether the pipeline fails or not will be analyzed.

1. If bias metrics exceed threshold, the entire pipeline fails.
2. If SHAP score drift from previous model is greater than 15%, the pipeline fails.
3. If critical features drop in importance, the pipeline fails.

A sample rule written for the above is:

```
if: bias.disparateImpact < 0.8
    fail_pipeline: true
```

8.6.1.5 Step 5 – Continuous Deployment (CD)

Once the model passes all the quality gates, it is deployed in SageMaker Endpoint.

8.6.1.6 Step 6 – Real-Time XAI for End Users

For instance, the loan sanctioning official, who is an end user of the above application, wants explanations through the UI as a query having the following feature values:

```
{
        "inputs": {
                "age": 45
                "income": 45000
                "gender": male
                "credit_score": 750
},
"explain": true
}
```

The output from the tool will contain:

- Predicted loan approval as Yes/No
- SHAP values for each feature to indicate its importance
- Explanation summary report

Sample Output:

```
{
    "prediction": "approved",
    "explanation":  {
           "credit_score": 0.69,
           "income": 0.25
           "age": -0.12
           "gender": -0.01
}}
```

On getting this explanation, the loan sanctioning official can understand which feature has the highest positive influence on the decision. It improves transparency in the decision-making process.

8.6.1.7 Step 7 – Continuous Monitoring

Amazon SageMaker Clarify monitors the following:

1. SHAP drift
2. Bias drift
3. Data distribution shift

For example, in the above case study, the tool will send alerts if "gender" bias increases, top features change significantly, or model behavior becomes unstable in the real-time environment.

8.7 RESPONSIBLE AI (RAI) VALIDATION USING AZURE ML

The following code is taken from "Microsoft Learn+" to understand how the model interpretability with RAI dashboard works.

8.7.1 Initial Set-Up Code

```
from azure.ai.ml import MLClient
from azure.identity import DefaultAzureCredential
from azure.ai.ml.entities import Model
from raiwidgets import RAIInsights
# auth + client
ml_client = MLClient(DefaultAzureCredential(), subscription_
  id, resource_group, workspace_name)
# assume you have: X_train, X_test, y_train, y_test, and
  trained_model (sklearn/onnx/pickle)
# Create an RAIInsights object (this is the notebook-based
  wrapper)
rai_insights = RAIInsights(clf=trained_model,
                X_train=X_train, y_train=y_train,
                X_test=X_test, y_test=y_test,
                feature_names=list(X_train.columns),
                categorical_features=['gender', 'region'])
# Compute explanations (SHAP), fairness, counterfactuals
rai_insights.explain()      # computes global/local
                              explanations
rai_insights.fairness()     # runs fairness metrics (wraps
                              Fairlearn etc.)
rai_insights.save('rai_insights.json')   # artifact to upload
# register model + attach RAI artifact (example pseudocode)
model = Model(path="model.pkl", name="<MODEL_NAME>",
  type="custom_model")
registered = ml_client.models.create_or_update(model)
# upload RAI artifact to storage and register link in model
  metadata
```

8.7.2 Example Azure DevOps Pipeline (YAML) – Key Stages

This is a simplified Azure Pipeline showing where XAI steps run. Replace az ml with azure-ai-ml CLI or calls to Python scripts.

```
trigger:
  branches: [ main ]
pool:
  vmImage: 'ubuntu-latest'
variables:
  AZ_SUBSCRIPTION: '<sub>'
```

```
  AZ_RESOURCE_GROUP: '<rg>'
  AZ_WORKSPACE: '<ws>'
stages:
- stage: Build
  jobs:
  - job: UnitTests
 steps:
    - script:
        pip install -r requirements.txt
        pytest tests/
      displayName: 'Run unit tests'
- stage: TrainAndExplain
  dependsOn: Build
  jobs:
  - job: TrainAndRAI
  steps:
    - script:
        pip install azure-ai-ml raiwidgets fairlearn shap
        python train_full_model.py --output model.pkl
        python run_rai_analysis.py --model model.pkl --out
          rai_bundle.json
      displayName: 'Train model & run Responsible AI analysis'
    - task: PublishBuildArtifacts@1
      inputs:
        pathToPublish: '$(Build.SourcesDirectory)/artifacts'
        artifactName: 'model-and-rai'
- stage: QualityGate
  dependsOn: TrainAndExplain
  jobs:
  - job: Evaluate
    steps:
    - script: |
        python evaluate_metrics.py --rai artifacts/rai_bundle.
          json --threshold_file gating_config.json
        # script should exit nonzero if gating fails
      displayName: 'Run gating rules for accuracy, fairness,
        explanation stability'
- stage: Deploy
  dependsOn: QualityGate
  condition: succeeded()
  jobs:
  - job: DeployToStaging
    steps:
    - script: |
        az ml model register -n <MODEL_NAME> -p model.pkl
        az ml online-endpoint create -f endpoint-staging.yml
        az ml online-deployment create -f deploy-staging.yml
           --all-traffic
      displayName: 'Deploy to staging endpoint'
```

8.7.3 Example Gating Logic (Pseudocode)

```
evaluate_metrics.py should enforce both performance and XAI
  checks, e.g.:
import json
rai = json.load(open('artifacts/rai_bundle.json'))
accuracy = rai['metrics']['accuracy']
disparate_impact = rai['metrics']['fairness']
  ['disparate_impact']
shap_corr = rai['metrics']['explainability']
  ['shap_correlation_with_baseline']
if accuracy < 0.85:
    raise SystemExit("FAIL: accuracy too low")
if disparate_impact < 0.8:
    raise SystemExit("FAIL: disparate impact too low (possible
       bias)")
if shap_corr < 0.8:
    raise SystemExit("FAIL: explanation stability failure")
print("PASS")
```

8.8 PRODUCTION MONITORING AND EXPLANATION DRIFT

Data drift detection: configure Azure ML Data Drift Monitor to continuously compare inference data distribution vs. training baseline; when drift passes threshold, trigger a RAI reanalysis batch job.

Explanation drift: schedule a job that recomputes global SHAP on a random sample of recent predictions and compares the feature-importance ordering/SHAP magnitudes to the baseline (compute Spearman or Pearson correlation).

Automated retrain: if drift + performance drop + explanation drift cross thresholds, trigger a retraining pipeline (with human-in-the-loop approval).

Dashboards & audits: save snapshot RAI dashboards for each model version for audit and compliance; Azure ML RAI Dashboard supports interactive exploration of fairness, error analysis, and feature importance.

Version everything: model, explained artifacts (SHAP arrays), fairness reports, and dataset snapshot – store references in model registry metadata. This makes audits reproducible. Choose baseline comparisons: a fixed baseline model/period is needed to compute explanation stability and drift.

Human review gates: for “soft” XAI failures (explanation changes but accuracy is fine), route to human reviewer rather than automatic rejection. Use Azure Boards/PR flow.

Scale explainability runs: full SHAP on very large datasets is expensive – run on representative samples for CI gating and run full runs in scheduled nightly jobs.

Mitigation library: integrate Fairlearn for mitigation experiments and record mitigated model candidates alongside original for decision makers.

8.9 SUMMARY

The CI/CD pipeline integrated with XAI brings the model to be retrained if there is a bias in the model's prediction in real-time environment. Thus, the various validation checks and the decisions derived based on them provide insight into the model's viability in the decision-making process. This chapter also showcases two different tools to address these concerns, which will definitely help the model builders and developers to better understand the way by which CI/CD pipeline will be integrated with XAI.

BIBLIOGRAPHY

Azure ML pipelines/model lifecycle (versioning, monitoring).

Azure DevOps CI/CD for Azure machine learning (pipelines & examples). Microsoft Learn. https://learn.microsoft.com/en-us/azure/machine-learning/how-to-responsible-ai-dashboard?view=azureml-api-2&utm_source=chatgpt.com

Machine learning fairness/Fairlearn integration in Azure ML. Microsoft Learn. https://learn.microsoft.com/en-us/azure/machine-learning/how-to-responsible-ai-dashboard?view=azureml-api-2&utm_source=chatgpt.com

Model interpretability in Azure ML (InterpretML/SHAP integrations). Microsoft Learn. https://learn.microsoft.com/en-us/azure/machine-learning/how-to-responsible-ai-dashboard?view=azureml-api-2&utm_source=chatgpt.com

Responsible AI dashboard – Azure machine learning. Microsoft Learn. https://learn.microsoft.com/en-us/azure/machine-learning/how-to-responsible-ai-dashboard?view=azureml-api-2&utm_source=chatgpt.com

9 XAI in DL Model Deployment, Monitoring, and Maintenance

9.1 INTRODUCTION

As AI systems are increasingly becoming a part of critical decision-making applications in domains such as healthcare, finance, transportation, and governance – the need for transparency, trustworthiness, accountability, and interpretability in deployed models has become paramount. XAI addresses this by providing mechanisms that make the behavior of the model, predictions made by the model, and decisions of AI systems more understandable to humans.

While development and training phases have seen significant progress in integrating XAI, the need for integrating it during the deployment phase remains a complex and essential challenge. It is essential to understand how this can be done before deploying Deep Learning models, which are black boxes in nature.

Subsequent sections will discuss how this integration can be achieved using real-time XAI model implementation for case studies.

9.2 NEED FOR EXPLAINABILITY IN DEPLOYMENT

Deployment of model is nothing but transitioning the developed and tested model to real-world operational settings. In this stage, models interact with live streams of data, handle different user bases, and operate in dynamic work environments.

Consequently, ensuring real-time interpretability and trustworthiness becomes crucial, as the stakeholders range from novice end users to domain experts. They all need insights on how the decisions have been made especially with respect to ethical and legal implications. Thus, integrating XAI in deployment phase ensures continuous transparency, model behavior monitoring, and compliance with AI governance standards such as GDPR's "right to explanation."

9.2.1 Architectural Integration of XAI in Deployment

XAI can be integrated into the deployment phase through the following strategies:

1. Explainability as a Service (EaaS): By means of deploying separate microservices that can generate explanations on demand, which can be accessed through APIs, it can be achieved. For instance, models that expose the endpoints not only for predictions but also for interpretability outputs such as SHAP- or LIME-based explanations.

 DOI: 10.1201/9781003630241-11

2. Explainability Layer Integration: By incorporating interpretable or explainable modules directly in the deployment pipeline, it ensures explanations are generated along with each inference or decision in real time.
3. Posthoc Dashboards and Monitoring Systems: By integrating visualization dashboards which can continuously analyze the model performance, feature importance changes, and decision rationales during production.

The architectures discussed earlier help developers and other stakeholders access the explanations without compromising scalability or latency issues, as they are more important when deployment is done at large scale.

9.3 ADVANTAGES OF XAI-DRIVEN MODEL DEPLOYMENT

The XAI integration in deployment pipelines provides the following benefits:

1. Trust and Accountability – It helps in building confidence among all the stakeholders by ensuring every model decision is interpretable.
2. Ethical Compliance – It enables adherence to explainability regulations and AI ethics frameworks.
3. Operational Insight – It assists DevOps and MLOps teams in diagnosing model errors, biases, and performance anomalies by means of transparent and interpretable reports.
4. Empowering End Users – It provides understandable reasoning behind automated recommendations by means of informed decision-making.

9.4 DEPLOYMENT PHASE WITH XAI INTEGRATION

With the rapid increase in the availability of large-scale image datasets and the parallel advancements in computational capabilities (e.g., powerful GPUs and cloud computing), CNNs have become the dominant solution for many real-world applications. These include autonomous driving systems, where object detection is critical for safety, medical imaging for diagnostic tasks, and even fields such as facial recognition, surveillance, and robotics. In particular, CNNs can perform these tasks efficiently and accurately because of their ability to learn from data with minimal human intervention.

However, as the model's working is a black box, it is very difficult to understand how the model has come up with a specific decision. Hence, an integrated workflow that combines XAI at the level of deployment is given below in Figure 9.1.

9.4.1 Model Architecture

Convolutional Neural Networks (CNNs) have revolutionized the field of computer vision and image classification by leveraging their hierarchical structure to automatically learn and extract features from images.

CNNs can learn the features that are impacting the outcome directly using the raw pixel data, which makes them a suitable solution for image-related applications.

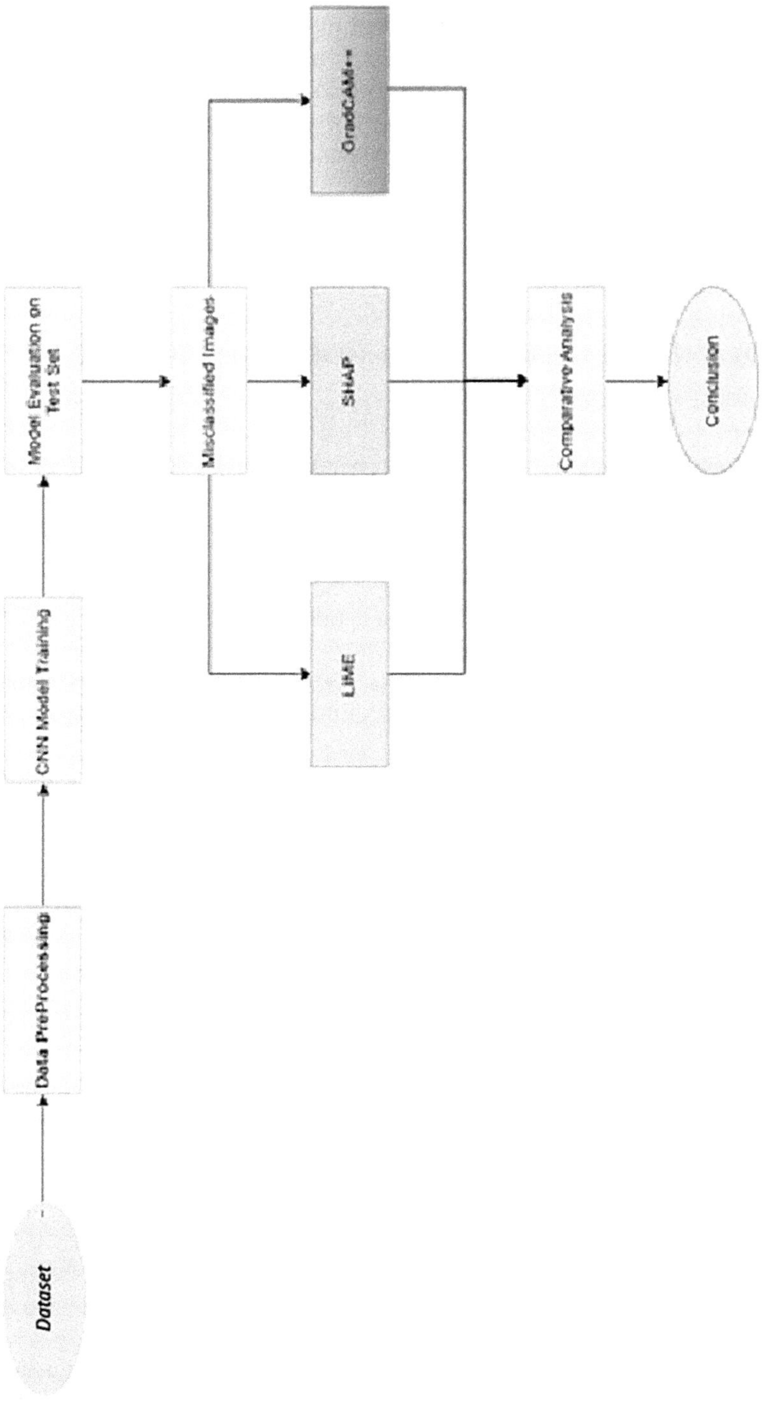

FIGURE 9.1 An integrated workflow with XAI at the level of deployment.

The CNN architecture contains multiple convolutional layers that perform convolution operations, activation functions, and pooling layers, which help the model get relevant hierarchies of the data.

In order to get a better understanding, this chapter first applies three prominent CNN models such as VGG16, VGG19, and ResNet50.

The CIFAR-10 dataset is used in all the models to find their performance results.

9.4.2 Application of VGG16 Model

9.4.2.1 Data Preprocessing

Under data preprocessing, several preprocessing functions were applied to get the dataset ready for classification.

9.4.2.2 Data Normalization

Data normalization is applied to normalize the pixel values within a certain range. Here, all the images in CIFAR-10 dataset are normalized to the range [0,1] to achieve convergence. The corresponding formula applied is given below:

$$x\_\text{norm} = x / 25$$

Here, x is the pixel intensity. This normalization process helps ensure all the features are of uniform scale. This improves the stability of the learning process.

9.4.2.3 Data Augmentation

In order to avoid overfitting, data augmentation techniques are applied. The techniques applied are: random horizontal flipping, small random rotations, zooming, and random translation.

This step is essential to improve model generalization so that the DL model can be applied on a wider range of input images for training.

Some of the sample parameter settings are:

Random Horizontal Flipping probability 0.5
Random rotation is between [-10,+10]
Zooming in the range [0.9,1.1]
Batch size is having mini-batches of size 32 and the training set was shuffled after every epoch to avoid biases.

The specific DL models are applied on this preprocessed dataset.

9.5 EXPLAINABILITY TECHNIQUES

Here, the XAI techniques such as LIME, SHAP and GradCAM++ are applied on each of these models.

9.5.1 LIME (Local Interpretable Model-Agnostic Explanations)

LIME works on the principle of perturbation. It generates perturbed samples of the input image by altering pixel values or applying slight transformations on it.

LIME then evaluates the DL model's prediction on these perturbed image samples and fits it to a local linear model is called a "surrogate model" for this data while focusing on the vicinity of the original instance. This linear model provides insights into the influencing features that impact the DL model's prediction.

From the CIFAR-10 dataset, LIME helps in finding which pixels in the images are contributing to the classification of an image by the DL model. This insight is provided by highlighting the important regions within the image. This is a local explanation as it provides visualization on the model's reasoning at a granular level.

9.5.2 SHAP (SHapley Additive exPlanations)

This technique provides a unified score for feature importance based on its impact on the prediction. The final score of the contribution will be equal to the model's prediction. These values are calculated using Kernel SHAP method. This method approximates the Shapley values by taking different subsets of the feature set as samples. It quantifies the contribution of each pixel in the classification, which allows the developers to visualize the most influential features for each prediction.

The SHAP plots generated provide a visual representation of feature importance, which helps in finding patterns and anomalies in the model's decision.

9.5.3 Gradient-Weighted Class Activation Mapping (GradCAM++)

The next XAI technique applied is GradCAM++. This technique visualizes the regions of the input image which are impacting the model's prediction. This is done using the gradients of the output class corresponding to the final convolutional layer of the DL model by highlighting the salient areas in the image that contributed to the decision.

From the gradients, a weighted average is calculated to get class activation maps. These maps are then overlaid on the original images to get a clear visualization of which regions of the image are focused on by the model for its prediction.

9.6 EVALUATION METRICS: FIDELITY, UNAMBIGUITY, AND INTERPRETABILITY

In order to assess the effectiveness of the XAI techniques, metrics such as fidelity, unambiguity, and interpretability size are calculated.

9.6.1 Fidelity

This metric assesses whether the model's predictions are aligned with the explanations derived from the XAI techniques. If the score is high, it means the explanations are reliable. It is given in Equation 9.1.

$$Fidelity = average(\rho) = \left(1/|X| * \Sigma i\, \rho\right) \tag{9.1}$$

9.6.2 Unambiguity

This metric measures the clarity and consistency of the explanations generated by the XAI techniques. The perception of unambiguity will enhance the usability of the interpretations given by the model.

The ambiguity for a feature I is calculated using the following formula:

$$C\_(i) = \max\_{\{j=1..M\}}\left[\mathrm{Pl}(\mathrm{I}) - \mathrm{Bel}(\mathrm{I_j})\right] \tag{9.2}$$

9.6.3 Interpretability Size

It finds the interpretability size of the explanations generated. The size should be average and providing impactful insights on the explanations derived from the XAI techniques.

It is calculated using the formulas given in Equations 9.3 and 9.4.

Stability and consistency as mentioned earlier is given by

$$S = 1/m\sum Var(Edi), \tag{9.3}$$

With these values at hand, we can now express the Interpretability score (IS) as

$$Interpretability\ score(IR) = \alpha * fidelity + \beta * (1 - S) + \gamma * C \tag{9.4}$$

9.7 CASE STUDY

9.7.1 VGG16 Model Results

The dataset is processed and analyzed by the VGG16 model. The training, testing, and validation accuracy-loss graphs are shown in Figure 9.2. The confusion matrix as shown in Figure 9.3 gives the true positive, false positive, true negative, and false negative values of the model based on its classification results on the given dataset. However, it is quite difficult to draw any conclusion on whether the model is trustworthy from these figures. The VGG-16 model's performance metrics values are given in Table 9.1.

9.7.2 Explainability Results

9.7.2.1 LIME Implementation

LIME (Local Interpretable Model-agnostic Explanations) was used to generate local explanations for individual predictions. By perturbing the input image, LIME identified which features influenced the model's decision.

For several misclassified images, LIME highlighted important areas that contributed to the model's predictions (see Figure 9.4).

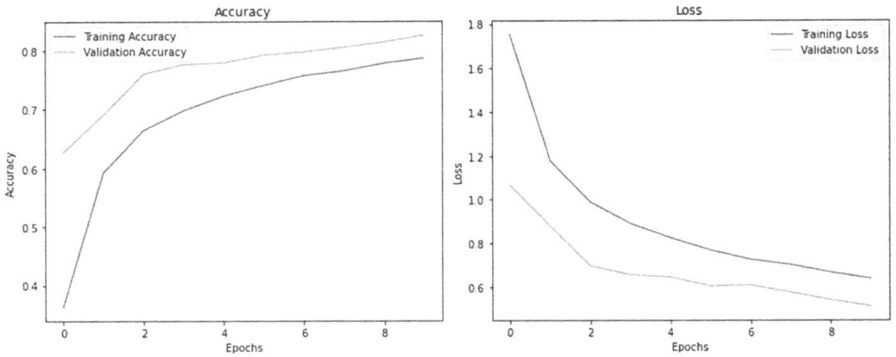

FIGURE 9.2 Accuracy-loss graph for the VGG16 model to illustrate the convergence pattern and the generalization gap between training and validation.

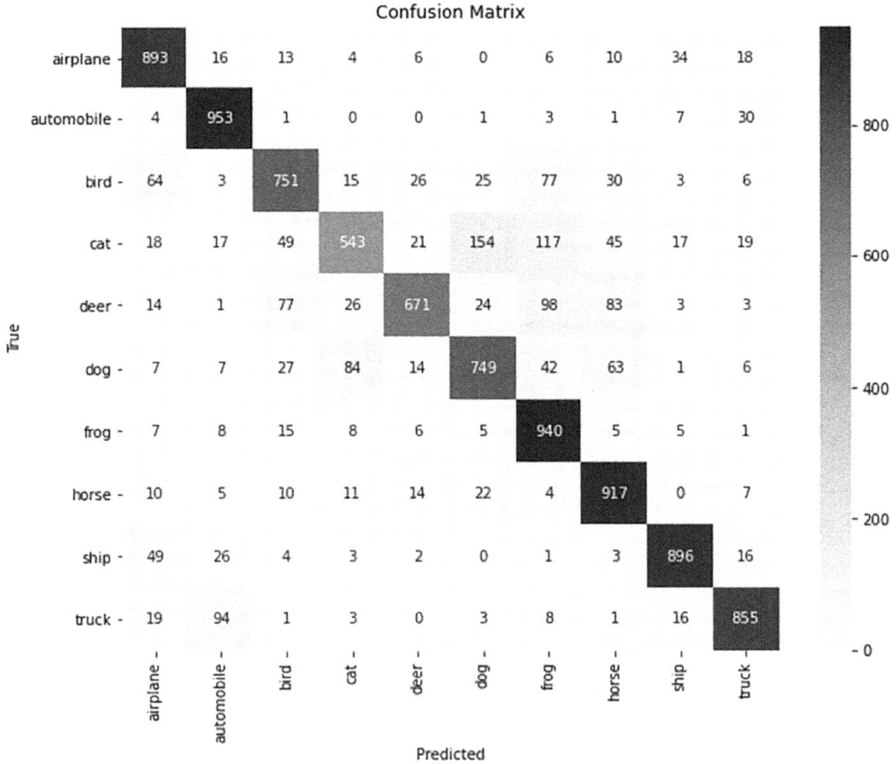

FIGURE 9.3 VGG16 confusion matrix to visually assess misclassification patterns, particularly among animal categories.

TABLE 9.1
VGG16 Model Evaluation Metrics

Model	Accuracy	Precision (Avg)	Recall (Avg)	F1-Score (Avg)	Best Class (F1-Score)	Challenging Class (F1-Score)
VGG16	82%	0.82	0.82	0.81	Ship (0.90)	Cat (0.64)

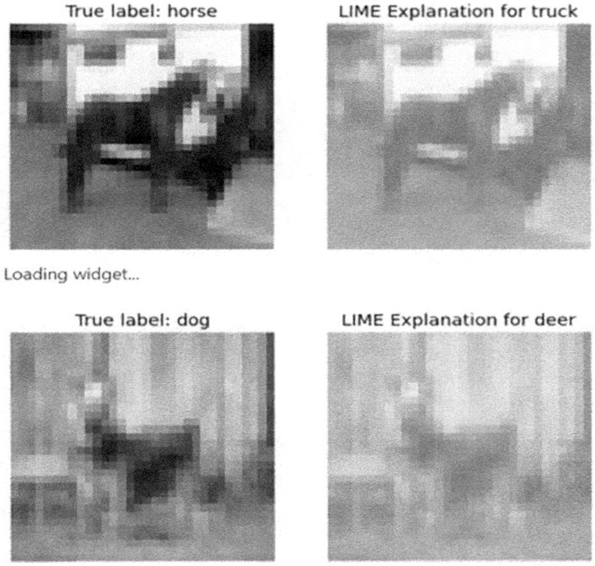

FIGURE 9.4 VGG16 model prediction with LIME explanation.

9.7.2.2 SHAP Implementation

The SHAP values are calculated on the set of misclassified images to find the features which have led to the wrong classification. The output provided from the SHAP plots indicates which pixels have higher influence on the decision (see Figure 9.5).

9.7.2.3 GradCAM++ Implementation

This technique analyzes the gradients of the output class obtained from the final convolutional layer. Then, the GradCAM++ activation maps are used to find the highlighted areas that influence the model's prediction results (see Figure 9.6).

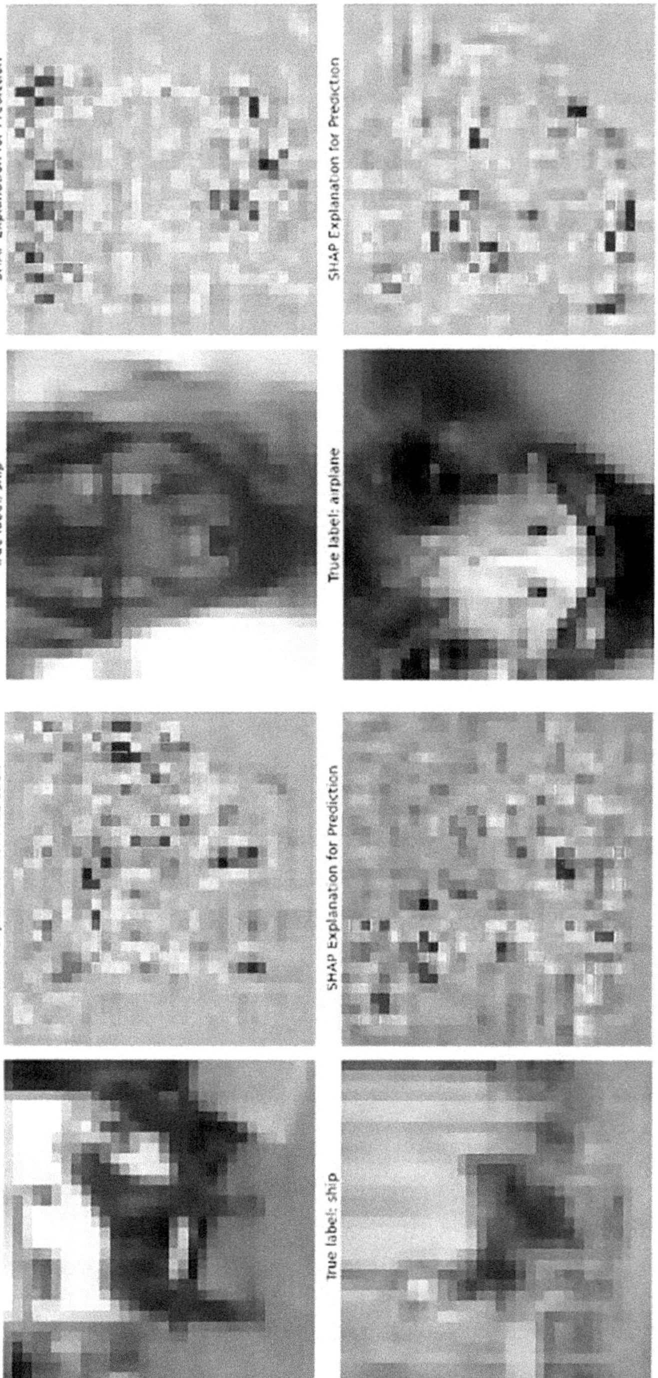

FIGURE 9.5 VGG16 misclassified images with SHAP explanation.

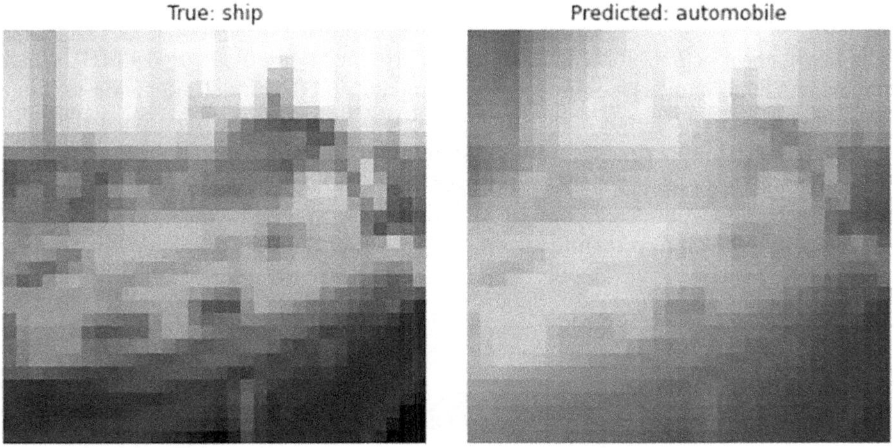

FIGURE 9.6 VGG16 prediction with GradCAM++ explanation.

9.8 EVALUATION OF EXPLAINABILITY TECHNIQUES

In this chapter, the VGG16 model is implemented with the evaluation of XAI methods such as LIME, SHAP, and GradCAM++. The explanations generated by the XAI methods are evaluated using metrics such as fidelity, unambiguity, and interpretability size. This will help us find which XAI method is suitable for VGG16 model to get reliable insights from the model's decision-making process.

Let us now discuss some of the insights derived from each of the applied XAI methods.

9.8.1 LIME

The LIME technique provided a fidelity score of 0.80, which means the model's decision-making is visualized by the LIME method with a higher level of trustworthiness. It aligns with the model's predictability.

It has an unambiguity score of 0.795, which shows it has a high level of clarity in the explanations generated, which reduces the complexity in understanding the explanations.

It has an interpretability size score of 0.77, which leads to the conclusion that the explanations are easily understandable in finding the importance of each feature in the model's decision-making.

9.8.2 SHAP

The application of SHAP on VGG16 model provides a fidelity score of 0.50, which means that it has some issues in fully capturing the VGG16 model's decision-making. This indicates SHAP is not a trustworthy solution for the model's explanations.

SHAP has a higher unambiguity score of 0.90, which means that the explanations are generated with higher clarity. However, as the fidelity score is very low, it leads to poor selection of the method.

In addition, the interpretability score is 0.65, which shows a moderate level of user accessibility and indicates the need for improvement in the interpretation of the XAI method.

9.8.3 GradCAM++

This XAI method has a fidelity score of 0.80, which provides a higher level of accuracy in local interpretability, which makes it effective for visual interpretations in VGG16 model.

The unambiguity score of 0.89 also provides a high level of support in generating clear explanations from the model without any complications.

In addition, it has an interpretability score of 0.73, which shows that this local visualization helps in understanding the VGG16 model's focus areas especially in image classification-based application domains (see Figure 9.7).

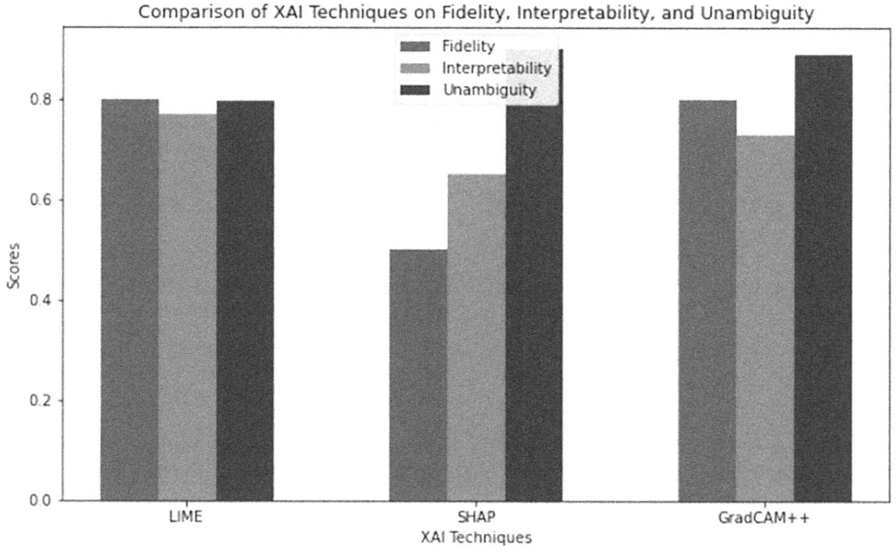

FIGURE 9.7 Explainability metrics evaluation.

9.9 SUMMARY

Deployment of model is nothing but transitioning the developed and tested model to real-world operational settings. In this stage, models interact with live streams of data, handle different user bases, and operate in dynamic work environments.

Consequently, ensuring real-time interpretability and trustworthiness becomes crucial, as the stakeholders range from novice end users to domain experts. They all need insights on how the decisions have been made especially with respect to ethical and legal implications. Thus, integrating XAI in deployment phase ensures continuous transparency, model behavior monitoring, and compliance with AI governance standards such as GDPR's "right to explanation."

The XAI techniques such as LIME, SHAP, and GradCAM++ are applied on each of these models. The various metrics applied to evaluate these XAI methods are: fidelity, unambiguity, and interpretablity. This helps in deciding which XAI method is more suitable for a high-performing model in a given application domain.

BIBLIOGRAPHY

Aldughayfiq, Bader, Ashfaq, Farzeen, Jhanjhi, N. Z., and Humayun, Mamoona. (2023). Explainable AI for retinoblastoma diagnosis: Interpreting deep learning models with LIME and SHAP. *Diagnostics*, 13(11),1932.

Ashraf, Kahakashan, Sadia Nawar, Md Hosen, Hamid, Islam, Mohammad Tanvirul, and Uddin, Mohammed Nazim. (2024). Beyond the black box: Employing LIME and SHAP for transparent health predictions with machine learning models. In *International Conference on Advances in Computing, Communication, Electrical, and Smart Systems (iCACCESS)*, Dhaka, Bangladesh, pp. 1–6. IEEE.

Begum, Momotaz, Shuvo, Mehedi Hasan, Nasir, Mostofa Kamal, Hossain, Amran, Hossain, Mohammad Jakir, Ashraf, Imran, Uddin, Jia, and Samad, Md Abdus. (2024). LCNN: Lightweight CNN architecture for software defect feature identification using Explainable AI. *IEEE Access*, 12, 55744–55756.

Bhandari, Mohan, Shahi, Tej Bahadur, Siku, Birat, and Neupane, Arjun. (2022). Explanatory classification of CXR images into COVID-19, pneumonia and tuberculosis using deep learning and XAI. *Computers in Biology and Medicine*, 150, 106156.

Bhandari, Mohan, Yogarajah, Pratheepan, Kavitha, Muthu Subash, and Condell, Joan. (2023). Exploring the capabilities of a lightweight CNN model in accurately identifying renal abnormalities: Cysts, stones, and tumors, using LIME and SHAP. *Applied Sciences*, 13(5), 3125.

Cervantes, Eduardo Gasca, and Chan, Wai-Yip. (2021). LIME-enabled investigation of convolutional neural network performances in COVID-19 chest X-ray detection. In *Proceedings of the IEEE Canadian Conference on Electrical and Computer Engineering (CCECE)*, pp. 1–6. IEEE.

Chaiwong, Niracha, Pannakkong, Warut, and Buddhakulsomsiri, Jirachai. (2021). Classifying saleability of lime by using convolutional neural network (CNN) approach. PhD diss., Thammasat University.

Farhood, Helia, Najafi, Mohammad, and Saberi, Morteza. (2023). Improving deep learning transparency: Leveraging the power of LIME heatmap. In *International Conference on Service-Oriented Computing*, Rome, Italy, pp. 72–83. Springer Nature Singapore, Singapore.

Gaur, Loveleen, Bhandari, Mohan, Razdan, Tanvi, Mallik, Saurav, and Zhao, Zhongming. (2022). Explanation-driven deep learning model for prediction of brain tumour status using MRI image data. *Frontiers in Genetics*, 13, 822666.

Hsu, Chung-Chian, Salahuddin Morsalin, S. M., Reyad, Md Faysal, and Shakib, Nazmus. (2023). Artificial intelligence model interpreting tools: SHAP, LIME, and anchor implementation in CNN model for hand gestures recognition. In *International Conference on Technologies and Applications of Artificial Intelligence*, Yunlin, Taiwan, pp. 16–29. Springer Nature Singapore, Singapore.

Nguyen, Hung Truong Thanh, Cao, Hung Quoc, Nguyen, Khang Vo Thanh, and Pham, Nguyen Dinh Khoi. (2021). Evaluation of explainable artificial intelligence: SHAP, LIME, and CAM. In *Proceedings of the FPT AI Conference*, Hanoi, Vietnam, pp. 1–6.

Rahnama, Arash, and Tseng, Andrew. (2021). An adversarial approach for explaining the predictions of deep neural networks. In *Proceedings of the IEEE/CVF Conference on Computer Vision and Pattern Recognition*, pp. 3253–3262.

Sangroya, Amit, Rastogi, Mouli, Anantaram, C., and Vig, Lovekesh. (2020). Guided-LIME: Structured sampling based hybrid approach towards explaining blackbox machine learning models. In *CIKM (Workshops)*.

Vimbi, Viswan, Shaffi, Noushath, and Mahmud, Mufti. (2024). Interpreting artificial intelligence models: A systematic review on the application of LIME and SHAP in Alzheimer's disease detection. *Brain Informatics*, 11(1), 10.

Part 3

Emerging Trends in XAI

10 Integrating XAI in Agentic AI Architecture

Subramaniam Ganesan and Rajkumar J. Bhojan

10.1 INTRODUCTION

In the case of "Agentic AI" systems, the AI-driven decisions taken by the agents are completely black box and lead to issues in terms of trustworthiness and fidelity. To get insight into the decision-making process of the agents, it is essential to get explanations on how each of the agents provides such decisions to the end users. Hence, XAI needs to be integrated in these agents in order to get interpretable justifications on how the decision is derived from the available information.

For instance, in healthcare applications, XAI methods such as LIME, SHAP, rule-based reasoning techniques, and counterfactual explanations on the agent systems will help in understanding the decision-making process of the internal algorithm design.

In this chapter, the case study taken is on a healthcare dataset and an agentic agent framework with XAI integration is applied to highlight how XAI helps in achieving a higher level of trustworthiness of the system along with the usefulness of the derived decisions.

Explainable AI (XAI) methods provide transparency in algorithm functionality by giving explanations in an abstract and understandable format to the stakeholders. As the autonomous agents have less human intervention, they need detailed explanations on how they work independently as well as collaboratively in such a complex real-time environment. This will improve the transparency and fidelity of the system developed.

10.2 EMBEDDING XAI IN AGENTIC AGENT ARCHITECTURE

XAI enables healthcare AI agents to provide clear, context-specific explanations for their diagnoses, treatment recommendations, and decision pathways, allowing medical professionals to understand, evaluate, and verify the system's reasoning. Implementing XAI in this context enhances not only safety and transparency but also supports regulatory compliance (e.g., with GDPR and FDA guidelines) and ethical alignment. Ultimately, explainability transforms agentic healthcare AI from a black-box assistant into a collaborative partner, fostering informed decision-making and sustained clinical trust.

10.3 XAI-EMBEDDED AGENTIC AGENT FRAMEWORK

Generally, an agentic agent framework contains the following tasks:

- Perception
- Repository
- Planning

DOI: 10.1201/9781003630241-13

- Reasoning
- Action Execution
- Learning
- Human-in-the-Loop
- Goal Assessment

When XAI is embedded in each of the above tasks, it becomes easier to find the reason behind each action taken by the tasks. Here, every internal step will be explainable and the output will be justified with traceable policy updates.

The XAI embedding makes the life cycle as:

Agent ➔ Explanation ➔ Action ➔ Explanation.

Now, the agentic agent framework embedded with XAI is provided in Figure 10.1 as having three layers:

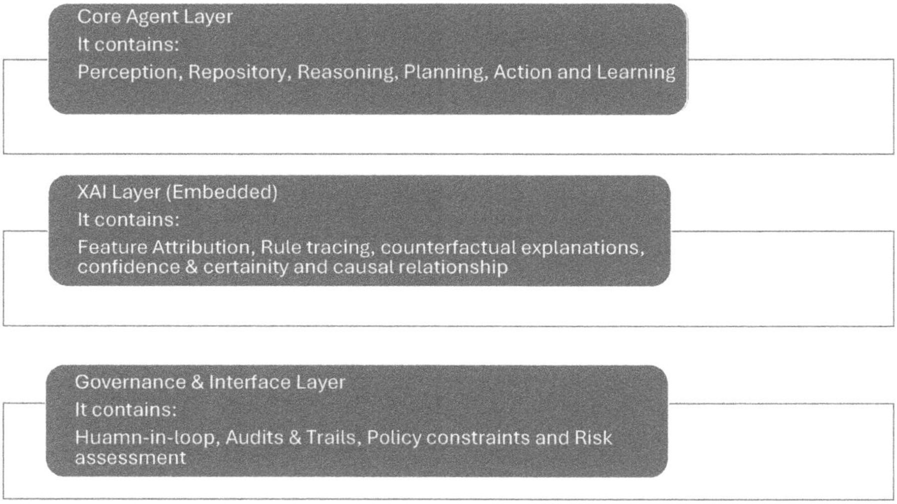

FIGURE 10.1 Agentic AI framework with XAI embedding.

10.4 BENEFITS OF IMPLEMENTING XAI IN AGENT WORK LIFECYCLE

If the agentic agent framework is embedded with XAI, it will surely improve the trustworthiness of the application. The benefits of each module embedded with XAI are given below:

1. Perception with XAI
   By applying XAI methods such as SHAP, Saliency Maps, etc., we can get feature attribution, attention visualization, sensor reliability-based visualization, and explanations. This will make the agent's world model, which is based on the perceptions, transparent and the input bias detectable.

2. Repository with XAI
   The agent will create, update, and use the repository of perceptions as a knowledge base to take decisions. It will be updated by means of previous experiences. The XAI embedding will help in achieving temporal traceability, detectable bias in the repository, and auditable experience updates.
3. Reasoning with XAI
   The major functionality of an agent is reasoning to choose actions and strategies based on perceptions. The embedding of XAI helps in providing rule tracing, identification of feature importance, uncertainty estimation, and policy explanations. This will lead to justifiable decisions, ethical compliance of decision, and autonomy validation.
4. Planning with XAI
   The agent has to provide a plan with subgoals to achieve. Embedding XAI in this stage helps in providing rationale in planning, comparison of other plans before confirming one, and explanations on previous plan failure.
   This will benefit easier replanning and human-in-the-loop-based alignment of planning to achieve long-term execution of the plan.
5. Action with XAI
   Based on the perceptions, reasoning, and planning, the agent has to take appropriate action in the environment in which it is deployed. XAI embedding at this stage helps in generating explanations on the risks associated with an action, expected outcome based on the action taken, etc. The benefits include accountability, prevention of unsafe actions, and approval of policy decisions.
6. Learning with XAI
   The agent learns through its experience from previous actions taken. This will update the model, the repository, and policies. XAI embedding will help in generating reward attribution, drift detection, and explanations on policy changes.

10.5 EXPERIMENTATION WITH XAI-EMBEDDED AGENTIC AGENT ARCHITECTURE

In our research, we focused on patient glucose-level monitoring. The XAI functionality was employed to interpret why alerts are triggered when glucose levels exceed the normal range, summarize trends and deviations from expected recovery patterns, and justify intervention recommendations. We utilized XAI tools such as LIME and SHAP in our experiments. Additionally, we used the Pima Indians Diabetes dataset and implemented the XGBoost model, which enabled interactive predictions with corresponding explanations.

Complete Implementation

```
!pip install lime
#Import libraries
import pandas as pd
import numpy as np
import matplotlib.pyplot as plt
```

```
import seaborn as sns
from sklearn.model_selection import train_test_split, cross_
   val_score, GridSearchCV
from sklearn.preprocessing import StandardScaler
from sklearn.metrics import classification_report, confusion_
   matrix, roc_auc_score, roc_curve
import xgboost as xgb
import warnings warnings.filterwarnings('ignore')
# XAI Libraries import shap import lime
from lime import lime_tabular
from typing import Union, List, Dict, Optional, Tuple
# Set style for better plots
plt.style.use('seaborn-v0_8')
sns.set_palette("husl")
print("All libraries imported successfully")
# Define the XAI Agent Class
class PimaDiabetesXAIAgent: XAI Prediction Agent for Pima
   Indians Diabetes Dataset using XGBoost
#This agent provides: Data loading and preprocessing XGBoost
   model training with hyperparameter tuning
#SHAP and LIME explanations
#Interactive predictions with explanations
#Model performance analysis
def   init  (self):
#Initialize the Pima Diabetes XAI Agent
self.model = None
self.scaler = StandardScaler()
self.X_train = None
self.X_test = None
self.y_train  =  None
self.y_test = None
self.feature_names = None
self.shap_explainer = None
self.lime_explainer = None
self.shap_values = None
# Feature escriptions for better interpretability
self.feature_descriptions = {
'Pregnancies': 'Number of times pregnant',
'Glucose': 'Plasma glucose concentration (2h oral glucose
   tolerance test)', 'BloodPressure': 'Diastolic blood
   pressure (mm Hg)',
'SkinThickness': 'Triceps skin fold thickness (mm)',
   'Insulin': '2-Hour serum insulin (mu U/ml)',
'BMI': 'Body mass index (weight in kg/(height in m)^2)',
   'DiabetesPedigreeFunction': 'Diabetes pedigree function
   (genetic factor)', 'Age': 'Age in years'
}
def load_data(self, data_source='url'):
# Load from a reliable source
url          =        "https://raw.githubusercontent.com/
   jbrownlee/Datasets/master/pima-indians-diabetes.data.csv"
```

```
column_names = ['Pregnancies', 'Glucose', 'BloodPressure',
   'SkinThickness',
'Insulin', 'BMI', 'DiabetesPedigreeFunction', 'Age',
   'Outcome']
try:
self.data = pd.read_csv(url, names=column_names)
print("Dataset loaded successfully from URL")
except Exception as e:
print("Error loading from URL: {e}")
print("Creating sample data for demonstration")
self._create_sample_data()
else:
try:
self.data = pd.read_csv('diabetes.csv')
print("Dataset loaded successfully from local file")
except FileNotFoundError:
print("Local file not found. Creating sample data for
   demonstration…")
self._create_sample_data()
self.feature_names = [col for col in self.data.columns if col
   != 'Outcome']
print("Dataset shape: {self.data.shape}")
print("Features: {self.feature_names}")
return self.data
def _create_sample_data(self):
#"Create sample data for demonstration purposes"
np.random.seed(42)
n_samples = 768
# Create realistic sample data based on Pima Indians Diabetes
   characteristics
data = {
'Pregnancies': np.random.poisson(3, n_samples),
'Glucose': np.random.normal(120, 30, n_samples).clip(0, 200),
'BloodPressure': np.random.normal(70, 15, n_samples).clip(0,
   120),
'SkinThickness': np.random.normal(20, 10, n_samples).clip(0,
   100),
'Insulin': np.random.normal(80, 100, n_samples).clip(0, 846),
'BMI': np.random.normal(32, 7, n_samples).clip(15, 70),
   'DiabetesPedigreeFunction': np.random.beta(2, 5, n_
   samples), 'Age': np.random.gamma(2, 10, n_samples).clip(21,
   81).astype(int)
}
# Create outcome based on realistic relationships
risk_score = ((data['Glucose'] > 140) * 2 +(data['BMI'] > 30)
   * 1.5 +(data['Age'] > 40) * 1 +
(data['Pregnancies'] > 5) * 0.5 +np.random.normal(0, 1,
   n_samples))
data['Outcome'] = (risk_score > 2).astype(int)
self.data = pd.DataFrame(data)
print("Sample data created for demonstration")
print("XAI Agent class defined successfully")
```

```
agent = PimaDiabetesXAIAgent()
data = agent.load_data()
# Display basic information about the dataset
print("\nDataset Preview:")
print(data.head())
print("\nDataset Info:")
print(data.info())
print("\nBasic Statistics:")
print(data.describe())
--output
Dataset loaded successfully from URL Dataset shape: (768, 9)
Features: ['Pregnancies', 'Glucose', 'BloodPressure',
   'SkinThickness', 'Insulin', 'BMI',
   'DiabetesPedigreeFunction
axes[0,0].pie(data['Outcome'].value_counts(), labels=['No
   Diabetes', 'Diabetes'], autopct='%1.1', startangle=90,
   colors=['lightblue', 'lightcoral'])
axes[0,0].set_title('Diabetes Distribution')
Text(0.5, 1.0, 'Diabetes Distribution')
```

Create Comprehensive EDA Visualizations

```
fig, axes = plt.subplots(2, 2, figsize=(15, 12))
# Feature correlation heatmap
corr_matrix = data.corr()
sns.heatmap(corr_matrix, annot=True, cmap='coolwarm',
   center=0, ax=axes[0,1], square=True, fmt='.2f')
axes[0,1].set_title('Feature Correlation Matrix')
# Age vs Glucose scatter plot colored by outcome
scatter  =  axes[1,0].scatter(data['Age'],  data['Glucose'],
c=data['Outcome'], cmap='viridis', alpha=0.6)
axes[1,0].set_xlabel('Age')
axes[1,0].set_ylabel('Glucose')
axes[1,0].set_title('Age vs Glucose (colored by Diabetes)')
plt.colorbar(scatter, ax=axes[1,0]) < matplotlib.colorbar.
   Colorbar at 0x7d9e439748c0>
#  axes[1,1].set_ylabel('Frequency')
# axes[1,1].set_title('BMI Distribution by Diabetes Status')
# axes[1,1].legend()
# plt.tight_layout()
# plt.show()
def preprocess_data(agent, test_size=0.2,
   scale_features=True):
print("Data Preprocessing")
# Handle missing values (zeros that should be NaN in medical
   context)
zero_cols = ['Glucose', 'BloodPressure', 'SkinThickness',
   'Insulin', 'BMI']
for col in zero_cols:
if col in agent.data.columns:
zero_count = (agent.data[col] == 0).sum() if zero_count > 0:
print("Replacing {zero_count} zeros in {col} with median")
agent.data[col] = agent.data[col].replace(0, agent.data[col].
   median())
```

```
# Separate features and target
X = agent.data[agent.feature_names]
y = agent.data['Outcome']
# Split data
agent.X_train, agent.X_test, agent.y_train, agent.y_test
   = train_test_split( X, y, test_size=test_size, random_
   state=42, stratify=y
)
# Scale features if requested if scale_features:
agent.X_train_scaled = pd.DataFrame( agent.scaler.fit_
   transform(agent.X_train),  columns=agent.feature_names,
   index=agent.X_train.index
)
agent.X_test_scaled = pd.DataFrame( agent.scaler.
   transform(agent.X_test),  columns=agent.feature_names,
   index=agent.X_test.index
)
else:
agent.X_train_scaled = agent.X_train.copy() agent.X_test_
   scaled = agent.X_test.copy()
print("Training set: {agent.X_train.shape}")
print("Test set: {agent.X_test.shape}")
print("Class distribution in training set: {agent.y_train.
   value_counts().to_dict()}")
# Preprocess the data
preprocess_data(agent)
#Data Preprocessing…
#Replacing 5 zeros in Glucose with median Replacing 35 zeros
   in BloodPressure with median Replacing #227 zeros in
   SkinThickness with median Replacing 374 zeros in Insulin
   with median Replacing 11 zeros #in BMI with median
Training set: (614, 8)
Test set: (154, 8)
Class distribution in training set: {0: 400, 1: 214}

def train_xgboost_model(agent, tune_hyperparameters=False):
print("Training XGBoost Model")
if tune_hyperparameters:
print("Performing hyperparameter tuning")
# Define parameter grid param_grid = {
'n_estimators': [100, 200],
'max_depth': [3, 4, 5],
'learning_rate': [0.1, 0.2],
'subsample': [0.8, 0.9],
'colsample_bytree': [0.8, 0.9]
}
# Grid search
xgb_model = xgb.XGBClassifier(random_state=42,
   eval_metric='logloss')
grid_search = GridSearchCV(
xgb_model, param_grid, cv=5, scoring='roc_auc', n_jobs=-1,
   verbose=1
)
```

```
grid_search.fit(agent.X_train_scaled, agent.y_train)
agent.model = grid_search.best_estimator_
print("Best parameters: {grid_search.best_params_}")
print("Best CV score: {grid_search.best_score_:.4f}")
else:
# Train with good default parameters agent.model = xgb.
   XGBClassifier(
n_estimators=100, max_depth=4, learning_rate=0.1,
   subsample=0.8, colsample_bytree=0.8, random_state=42,
   eval_metric='logloss'
)
agent.model.fit(agent.X_train_scaled, agent.y_train)
# Model evaluation
train_pred = agent.model.predict(agent.X_train_scaled)
test_pred = agent.model.predict(agent.X_test_scaled)
train_auc = roc_auc_score(agent.y_train, agent.model.
   predict_proba(agent.X_train_scaled)[:, 1]) test_auc
   = roc_auc_score(agent.y_test, agent.model.predict_
   proba(agent.X_test_scaled)[:, 1])
print("\n Model Performance:")
print("Training AUC: {train_auc:.4f}")
print(f" Test AUC: {test_auc:.4f}")
# Cross-validation scores
cv_scores = cross_val_score(agent.model, agent.X_train_scaled,
   agent.y_train, cv=5, scoring='roc_auc')
print(" CV AUC: {cv_scores.mean():.4f} (+/- {cv_scores.std() *
   2:.4f})")
return agent.model

# Train the model
model = train_xgboost_model(agent, tune_hyperparameters=False)
--output
Training XGBoost Model
Model Performance: Training AUC: 0.9888
Test AUC: 0.8281
# Create detailed performance visualizations fig,
axes = plt.subplots(2, 2, figsize=(15, 10))

# ROC Curve
y_pred_proba = agent.model.predict_proba(agent.X_test_scaled)
   [:, 1] fpr, tpr, _= roc_curve(agent.y_test, y_pred_proba)
roc_auc = roc_auc_score(agent.y_test, y_pred_proba)
axes[0,0].plot(fpr, tpr, color='darkorange', lw=2, label=f'ROC
   curve (AUC = {roc_auc:.3f})') axes[0,0].plot([0, 1], [0,
   1], color='navy', lw=2, linestyle='--')
axes[0,0].set_xlim([0.0, 1.0])
axes[0,0].set_ylim([0.0, 1.05])
axes[0,0].set_xlabel('False Positive Rate')
axes[0,0].set_ylabel('True Positive Rate')
axes[0,0].set_title('ROC Curve')
axes[0,0].legend(loc="lower right")
```

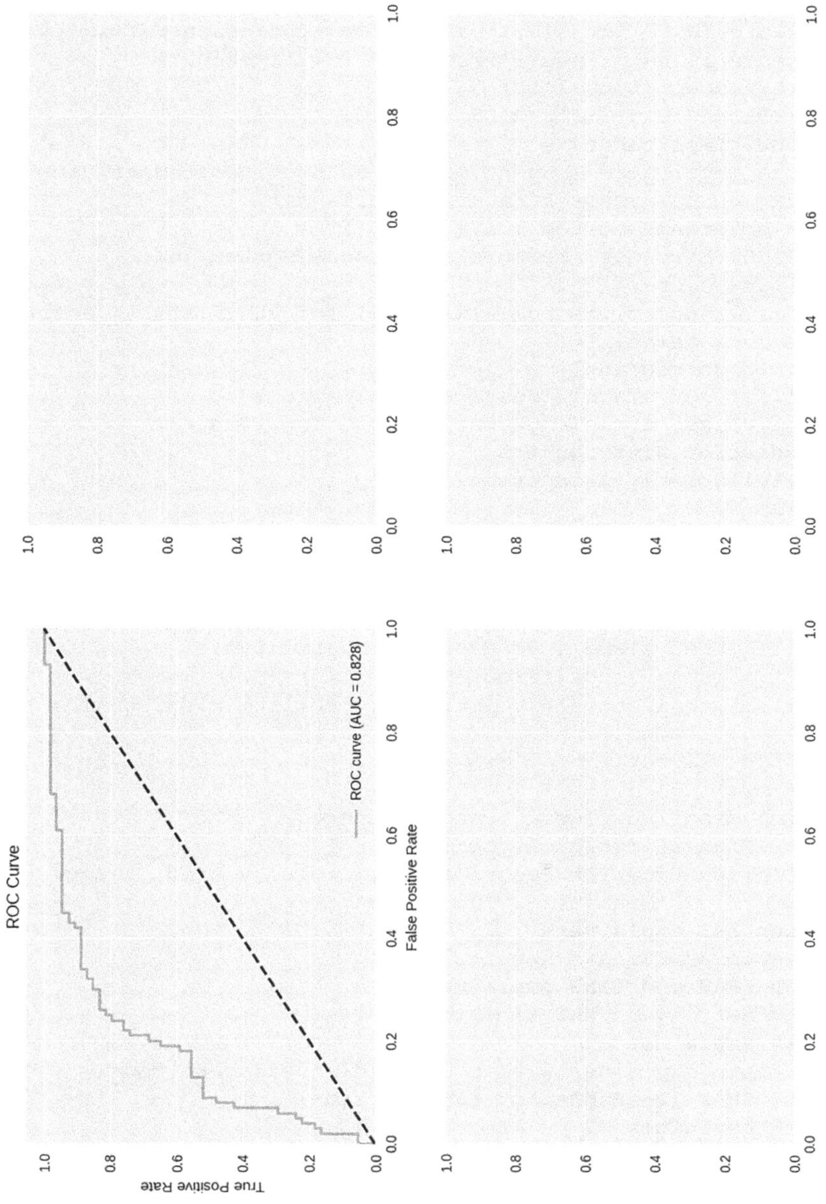
ROC Curve
ROC curve (AUC = 0.828)
False Positive Rate
True Positive Rate
0.0
0.2
0.4
0.6
0.8
1.0

```
# 2. Confusion Matrix

y_pred  =  agent.model.predict(agent.X_test_scaled) cm =
   confusion_matrix(agent.y_test, y_pred)
sns.heatmap(cm, annot=True, fmt='d', cmap='Blues',
   ax=axes[0,1]) axes[0,1].set_title('Confusion Matrix')
   axes[0,1].set_xlabel('Predicted')    axes[0,1].
   set_ylabel('Actual')
Text(792.3131313131312, 0.5, 'Actual')
#Feature Importance
feature_importance = agent.model.feature_importances_ sorted_
   idx  =  np.argsort(feature_importance)
pos = np.arange(sorted_idx.shape[0]) + .5
axes[1,0].barh(pos, feature_importance[sorted_idx],
   align='center')
axes[1,0].set_yticks(pos) axes[1,0].set_yticklabels([agent.
   feature_names[i]
for i in sorted_idx])
axes[1,0].set_xlabel('Feature Importance')
axes[1,0].set_title('XGBoost Feature Importance')
# Prediction Distribution
axes[1,1].hist(y_pred_proba[agent.y_test == 0], alpha=0.7,
   bins=20,
label='No Diabetes', color='blue')
axes[1,1].hist(y_pred_proba[agent.y_test == 1], alpha=0.7,
   bins=20,
label='Diabetes', color='red')
axes[1,1].set_xlabel('Predicted Probability')
axes[1,1].set_ylabel('Frequency')
axes[1,1].set_title('Prediction Probability Distribution')
axes[1,1].legend()
plt.tight_layout()
plt.show()
# Print detailed classification report
print("Classification Report:")
print(classification_report(agent.y_test, y_pred, target_
   names=['No Diabetes', 'Diabetes']))
# Setup XAI Explainers
def setup_explainers(agent):
#Setup SHAP and LIME explainers
print("Setting up XAI Explainers…")
# SHAP explainer
agent.shap_explainer = shap.TreeExplainer(agent.model)
print("SHAP TreeExplainer initialized")
# LIME explainer
agent.lime_explainer = lime_tabular.LimeTabularExplainer(
   agent.X_train_scaled.values,  feature_names=agent.
   feature_names,class_names=['No Diabetes', 'Diabetes'],
   mode='classification', discretize_continuous=True)
print("LIME TabularExplainer initialized")
```

```
# Compute SHAP values for test set
agent.shap_values = agent.shap_explainer.
   shap_values(agent.X_test_scaled)
print("SHAP values computed for test set")
# Setup the explainers
setup_explainers(agent)
# SHAP Summary Plots
print("Creating SHAP visualizations")
# Summary plot (dot plot)
plt.figure(figsize=(10, 6))
shap.summary_plot(agent.shap_values,agent.X_test_
   scaled,feature_names=agent.feature_names, show=False)
plt.title('SHAP Summary Plot - Feature Impact on Predictions')
plt.tight_layout()
plt.show()
# SHAP Summary plot (bar plot)
plt.figure(figsize=(10, 6))
shap.summary_plot(agent.shap_values,agent.X_test_scaled,
   feature_names=agent.feature_names, plot_type="bar",
   show=False)
plt.title('SHAP Feature Importance - Average Impact')
plt.tight_layout()
plt.show()
# SHAP Waterfall plot for a specific instance
sample_idx = 0
plt.figure(figsize=(10, 8))
```

```
Creating SHAP visualizations…
```

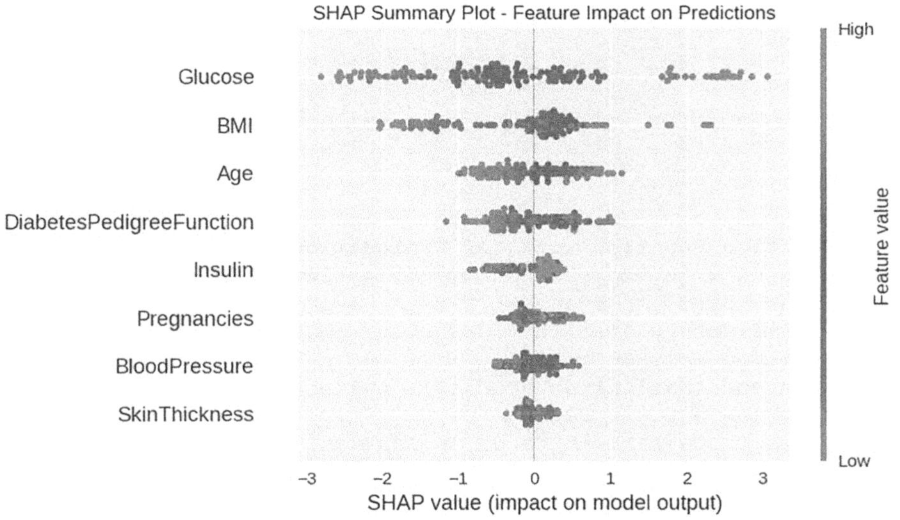

```
# Create explanation object for waterfall plot
explanation = shap.Explanation(values=agent.shap_
   values[sample_idx], base_values=agent.shap_explainer.
   expected_value, data=agent.X_test_scaled.iloc[sample_idx].
   values,  feature_names=agent.feature_names
)
shap.waterfall_plot(explanation, show=False)
plt.title(f'SHAP Waterfall Plot - Test Sample {sample_idx}')
plt.tight_layout()
plt.show()
print("Actual outcome for sample {sample_idx}: {agent.y_test.
   iloc[sample_idx]}")
print("Predictedprobability: {agent.model.predict_
   proba(agent.X_test_scaled.iloc[[sample_idx]])[0][1]:.3f}")
```

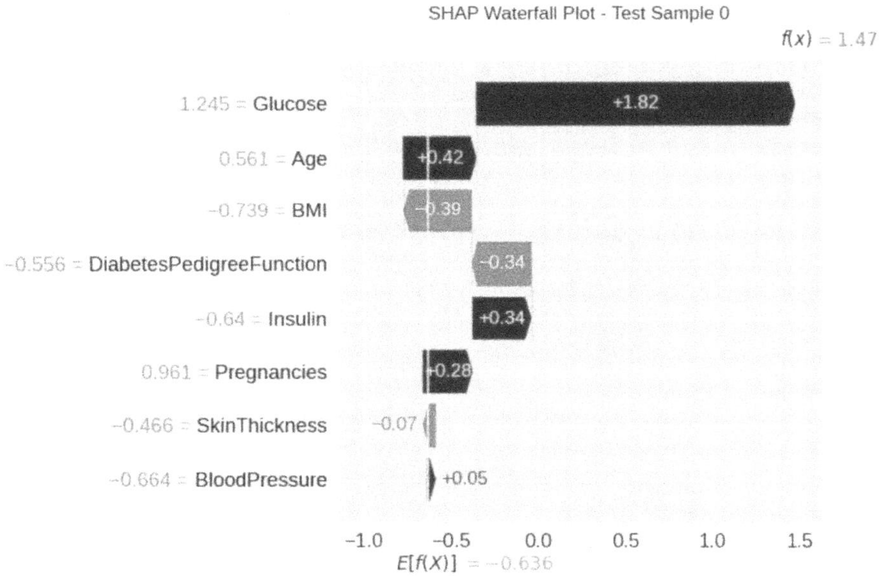


```
Actual outcome for sample 0: 0
```

```
# Prediction Function with XAI Explanations
def predict_with_explanation(agent, patient_data: Union[pd.
   Series, dict, np.ndarray],
patient_id: str = "Patient") -> Dict: "Make prediction with
   detailed explanation"
if isinstance(patient_data, dict): patient_data =
   pd.Series(patient_data)
elif isinstance(patient_data, np.ndarray):
patient_data  =  pd.Series(patient_data,  index=agent.
   feature_names)
# Scale the input
patient_scaled = pd.DataFrame([patient_data], columns=agent.
   feature_names)
```

```
patient_scaled = pd.DataFrame(
agent.scaler.transform(patient_scaled), columns=agent.
   feature_names
)
# Make prediction
prediction_proba = agent.model.predict_proba(patient_scaled)
   [0]
prediction  =  agent.model.predict(patient_scaled)[0]
confidence = prediction_proba[1]
# SHAP explanation
shap_values  =  agent.shap_explainer.
   shap_values(patient_scaled)
shap_explanation  =  dict(zip(agent.feature_names,
   shap_values[0]))
# LIME explanation
lime_exp = agent.lime_explainer.explain_instance( patient_
   scaled.iloc[0].values,  agent.model.predict_proba, num_
   features=len(agent.feature_names)
)
lime_explanation = dict(lime_exp.as_list())
# Risk assessment
if confidence < 0.3: risk_level = "Low"
elif confidence < 0.7: risk_level = "Moderate"
else:
risk_level = "High"
return {
'patient_id':  patient_id, 'prediction': int(prediction),
   'confidence': float(confidence), 'risk_level': risk_level,
   'patient_data': patient_data.to_dict(),
Shap_explanation': shap_explanation, 'lime_explanation':
   lime_explanation, 'feature_descriptions': agent.
   feature_descriptions
}
def visualize_explanation(explanation_result: Dict) -> plt.
   Figure: "Create visualization of the explanation"
fig, axes = plt.subplots(2, 2, figsize=(16, 12))
patient_id = explanation_result['patient_id']
#Prediction summary ax = axes[0, 0]
probabilities = [1-explanation_result['confidence'],
   explanation_result['confidence']]
colors = ['lightgreen', 'lightcoral']
bars = ax.bar(['No Diabetes', 'Diabetes'], probabilities,
   color=colors)
ax.set_title(f'{patient_id} - Diabetes Risk
   Prediction', fontsize=14, fontweight='bold')
   ax.set_ylabel('Probability')
ax.set_ylim(0, 1)
# Add probability labels on bars
for bar, prob in zip(bars, probabilities):
height = bar.get_height()
```

```
ax.text(bar.get_x() + bar.get_width()/2., height + 0.02,
   f'{prob:.2%}', ha='center', va='bottom', fontweight='bold')
# Add text with prediction details
ax.text(0.5, 0.8, f"Risk Level: {explanation_result['risk_
   level']}", transform=ax.transAxes, ha='center',
   fontsize=12, weight='bold', bbox=dict(boxstyle="round,
   pad=0.3", facecolor="lightblue"))
# SHAP explanation
shap_data  =  explanation_result['shap_explanation']
sorted_shap = sorted(shap_data.items(), key=lambda x:
   abs(x[1]), reverse=True)[:8]
features, values = zip(*sorted_shap)
colors_shap = ['red' if v < 0 else 'darkblue' for v in values]
ax = axes[0, 1]
bars = ax.barh(range(len(features)), values, color=colors_
   shap, alpha=0.7) ax.set_yticks(range(len(features)))
ax.set_yticklabels(features)
ax.set_xlabel('SHAP Value (Impact on Prediction)')
ax.set_title('SHAP Feature Importance', fontweight='bold')
   ax.axvline(x=0, color='black', linestyle='-', alpha=0.3)
# Add value labels
for i, (bar, value) in enumerate(zip(bars, values)):
ax.text(value + (0.01 if value >= 0 else -0.01), i,
   f'{value:.3f}', ha='left' if value >= 0 else 'right',
   va='center', fontsize=9)
# LIME explanation
lime_data  =  explanation_result['lime_explanation']
sorted_lime = sorted(lime_data.items(), key=lambda x:
   abs(x[1]), reverse=True)[:8]
features_lime, values_lime = zip(*sorted_lime)
colors_lime = ['red' if v < 0 else 'darkgreen' for v in
   values_lime]
ax = axes[1, 0]
bars = ax.barh(range(len(features_lime)),
   values_lime, color=colors_lime, alpha=0.7)
   ax.set_yticks(range(len(features_lime)))
ax.set_yticklabels(features_lime) ax.set_xlabel('LIME Value
   (Local Importance)')
ax.set_title('LIME Feature Importance', fontweight='bold')
ax.axvline(x=0, color='black', linestyle='-', alpha=0.3)
# Add value labels
for i, (bar, value) in enumerate(zip(bars, values_lime)):
ax.text(value + (0.01 if value >= 0 else -0.01), i,
   f'{value:.3f}', ha='left' if value >= 0 else 'right',
   va='center', fontsize=9)
# Patient data visualization
ax = axes[1, 1]
patient_data  =  explanation_result['patient_data']
# Create normalized patient data for better visualization
normalized_data = {}
ranges = {
```

```
'Pregnancies': (0, 15), 'Glucose': (50, 200), 'BloodPressure':
   (40, 120),
'SkinThickness': (0, 50), 'Insulin': (0, 300), 'BMI': (15,
   50),
'DiabetesPedigreeFunction': (0, 2), 'Age': (20, 80)
# Sample Predictions with Explanations
# Create sample patients for demonstration
sample_patients = {
"High Risk Patient": { 'Pregnancies': 8,
'Glucose': 183,
'BloodPressure': 64,
'SkinThickness': 23,
'Insulin': 672,
'BMI': 32.9,
'DiabetesPedigreeFunction': 0.672,
'Age': 50
},
"Low Risk Patient": { 'Pregnancies': 1,
'Glucose': 85,
'BloodPressure': 66,
'SkinThickness': 29,
'Insulin': 26,
'BMI': 26.6,
'DiabetesPedigreeFunction': 0.351,
'Age': 31
},
"Moderate Risk Patient": { 'Pregnancies': 4,
'Glucose': 110,
'BloodPressure': 92,
'SkinThickness': 15,
'Insulin': 125,
'BMI': 37.6,
'DiabetesPedigreeFunction': 0.191,
'Age': 30
}
}
# Make predictions and create explanations
for patient_name, patient_data in sample_patients.items():
   print("\n{'='60}")
print("ANALYSIS FOR {patient_name.upper()}")
 print("{'='60}")
# Get prediction with explanation
result = predict_with_explanation(agent, patient_data,
   patient_name)
# Print summary
print("Patient Data: {patient_data}")
print("\nPrediction: {'DIABETES RISK' if result['prediction']
   else NO DIABETES'}") print("Confidence:
   {result['confidence']:.2%}")
print("Risk Level: {result['risk_level']}")
# Top 3 most important features from SHAP
```

```
shap_sorted = sorted(result['shap_explanation'].items(),
key=lambda x: abs(x[1]), reverse=True)[:3]
print("\nTop Contributing Factors (SHAP):")
for i, (feature, importance) in enumerate(shap_sorted, 1):
direction = "increases" if importance > 0 else "decreases"
print(" {i}. {feature}: {direction} diabetes risk (impact:
   {importance:.3f})")
# Create visualization
fig = visualize_explanation(result) plt.show()
print("\n" + "-"*60)
```

```
#ANALYSIS FOR HIGH RISK PATIENT
Patient Data: {'Pregnancies': 8, 'Glucose': 183,
   'BloodPressure': 64, 'SkinThickness': 23, 'Insulin': 672,
   'BMI': 3
Prediction: DIABETES RISK Confidence: 92.51%
Risk Level: High
Top Contributing Factors (SHAP):
Glucose: increases diabetes risk (impact: 2.035)
Age: increases diabetes risk (impact: 0.535)
Insulin: decreases diabetes risk (impact: -0.426)
# Interactive Prediction Interface
#ANALYSIS FOR LOW RISK PATIENT
'Glucose': 85, 'BloodPressure': 66, 'SkinThickness': 29,
   'Insulin': 26, 'BMI': 26.
"Interactive function for making new predictions"
Prediction: NO DIABETES Confipdreincte("2.0D6I%ABETES RISK
   ASSESSMENT TOOL")
RiskpLreivnetl("=Low50")
print("Enter patient information for diabetes risk
   prediction:")
Glucose: decreases diabetes risk (impact: -2.135)
#BMC: decrases diabetes risk (impact: -1.190)
Pregnancies: decreases diabetes risk (impact: -0.218)
patient_data = {}
# Default values for demonstration defaults = {
'Pregnancies': 2,
'Glucose': 120,
'BloodPressure': 70,
'SkinThickness': 20,
'Insulin': 94,
'BMI': 28.7,
'DiabetesPedigreeFunction': 0.259,
'Age': 35
}
for feature in agent.feature_names:
description = agent.feature_descriptions[feature]
default_val = defaults[feature]
try:
user_input = input("{feature} ({description}) [{default_val}]:
   ").strip()
```

```
if user_input == "":
patient_data[feature] = default_val
else:
patient_data[feature] = float(user_input) except ValueError:
print("Invalid input, using default value: {default_val}")
patient_data[feature] = default_val
# Make prediction
result = predict_with_explanation(agent, patient_data,
   "Interactive Patient")
print("\n{'='*60}")
print("DIABETES RISK ASSESSMENT RESULTS")
print("{'='*60}")
print("Patient Data: {patient_data}")
printf("\nPrediction: {'HIGH DIABETES RISK' if
   result['prediction'] else 'LOW DIABETES RISK'}")
printf("Confidence: {result['confidence']:.2%}")
printf("Risk Level: {result['risk_level']}")
# Create visualization
fig = visualize_explanation(demo_result)
plt.show()
```

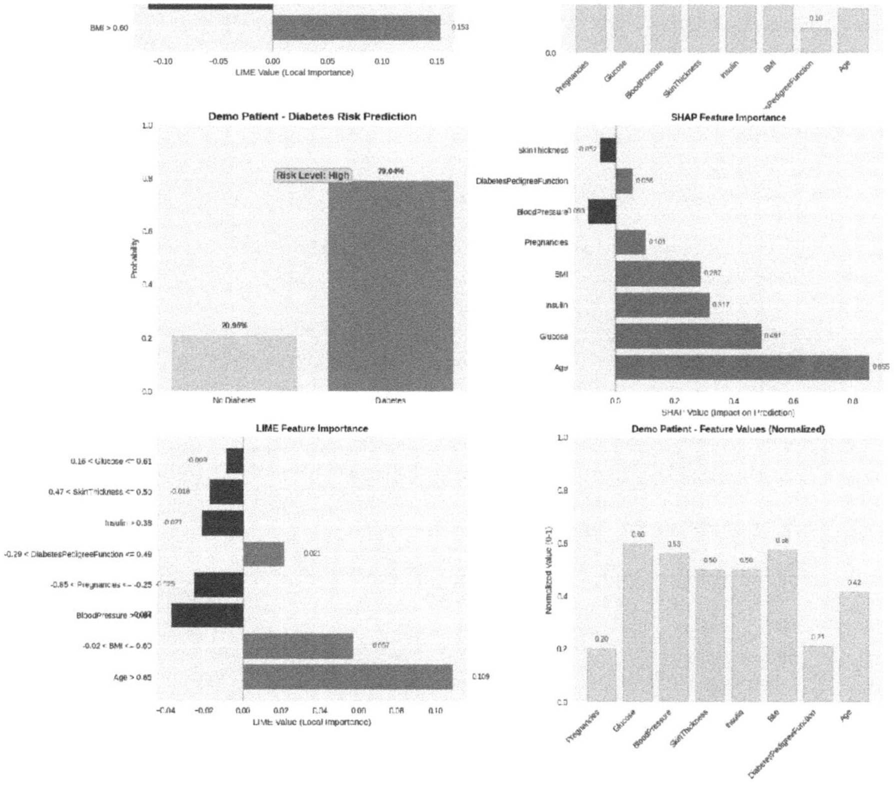

```
Patient Data: {'Pregnancies': 3, 'Glucose': 140,
   'BloodPressure': 85, 'SkinThickness': 25, 'Insulin': 150,
   'BMI': 3
Prediction: DIABETES RISK Confidence: 79.04%
Risk Level: High
axes[0].set_yticklabels(features)
axes[0].set_xlabel('Importance')
axes[0].set_title('XGBoost Feature Importance')
# SHAP importance
sorted_shap = sorted(shap_importance.items(), key=lambda x:
   x[1], reverse=True) features, values = zip(*sorted_shap)
axes[1].barh(range(len(features)), values, color='lightgreen')
axes[1].set_yticks(range(len(features))) axes[1].
   set_yticklabels(features)
axes[1].set_xlabel('Mean |SHAP Value|') axes[1].set_
   title('SHAP Global Importance')
# Correlation importance
sorted_corr = sorted(correlations.items(), key=lambda x: x[1],
   reverse=True) features, values = zip(*sorted_corr)
axes[2].barh(range(len(features)), values, color='lightcoral')
axes[2].set_yticks(range(len(features)))
axes[2].set_yticklabels(features)
axes[2].set_xlabel('|Correlation with Target|')
axes[2].set_title('Feature-Target Correlation')
plt.tight_layout()
plt.show()
return xgb_importance, shap_importance, correlations xgb_imp,
   shap_imp, corr_imp = compare_feature_importance()
# Print ranking comparison
print("\nFEATURE IMPORTANCE RANKINGS:")
print("="60)
print("{'Rank':<4} {'XGBoost':<25} {'SHAP':<25}
   {'Correlation':<25}")
print("-"60)
xgb_ranking = [item[0] for item in sorted(xgb_imp.items(),
   key=lambda x: x[1], reverse=True)] shap_ranking = [item[0]
   for item in sorted(shap_imp.items(), key=lambda x: x[1],
   reverse=True)] corr_ranking = [item[0] for item in
   sorted(corr_imp.items(), key=lambda x: x[1], reverse=True)]
for i in range(len(agent.feature_names)):
......
#FEATURE IMPORTANCE RANKINGS
```

1	Glucose	Glucose	Glucose
2	BMI	BMI	BMI
3	Age	Age	Age
4	Insulin	Diabetes PedigreeFunction	Pregnancies
5	Diabetes PedigreeFunction	Insulin	SkinThickness
6	SkinThickness	Pregnancies	Diabetes PedigreeFunction

10.6 SUMMARY

An agentic agent framework with XAI embedding integrates explanation mechanisms into every functional module of the framework. The modules such as perception, repository, reasoning, planning, action, and learning are embedded with XAI to achieve transparency on what the agent perceives, how it takes decisions, how it plans, which action it takes for a particular environmental perception, and how it learns from previous experience.

Such integration of XAI in the agentic agent framework transforms an agent's black-box decision-making into a transparent, accountable, and human-understandable system, which improves the safety, error correction, and reliability of the agent.

BIBLIOGRAPHY

Chakrabarty, P. K. (2025). Causal inference in Agentic AI: Bridging explainability and dynamic decision making. *International Journal of Science and Research (IJSR)*, 14(4), 10–21275.

Ciatto, G., Schumacher, M. I., Omicini, A., and Calvaresi, D. (2020, May). Agent-based explanations in AI: Towards an abstract framework. In *International Workshop on Explainable, Transparent Autonomous Agents and Multi-Agent Systems*, pp. 3–20. Springer International Publishing, Cham.

Desai, S. (2025). Agentic AI frameworks: Building autonomous, self-healing systems for financial infrastructure. *Journal of Computer Science and Technology Studies*, 7(12), 364–383.

Garg, V. (2025). Designing the mind: How agentic frameworks are shaping the future of AI behavior. *Journal of Computer Science and Technology Studies*, 7(5), 182–193.

Guasch, A., and Valdez, M. I. (2025). The STAR-XAI protocol: A framework for inducing and verifying agency, reasoning, and reliability in AI agents. arXiv preprint arXiv:2509.17978.

Kim, J., Mun, S., Lee, S., Cho, J., and Ok, J. (2025). Towards faithful agentic XAI: A verification method and an open-world benchmark for better. arXiv preprint arXiv:2605.27879.

Kubam, C. S. (2025). Agentic AI for autonomous, explainable, and real-time credit risk decision-making. arXiv preprint arXiv:2601.00818.

Silva, A., Schrum, M., Hedlund-Botti, E., Gopalan, N., and Gombolay, M. (2023). Explainable artificial intelligence: Evaluating the objective and subjective impacts of XAI on human-agent interaction. *International Journal of Human–Computer Interaction*, 39(7), 1390–1404.

Yamaguchi, T., Zhou, Y., Ryo, M., and Katsura, K. (2025). Agentic explainable artificial intelligence (Agentic XAI) approach to explore better explanation. arXiv preprint arXiv:2512.21066.

11 Interplay of XAI with GenAI

11.1 INTRODUCTION

LLMs are used in almost all applications to provide efficient communication to various stakeholders with an appropriate level of detail. Especially in healthcare analytics applications, the decision-making, communication with stakeholders, and documentation of medical transcripts have improved tremendously using LLMs.

As these models are capable of reading and writing like humans, their contextually appropriate responses and decision accuracy have led to their utilization in highly critical healthcare applications with improved precision in documentation processes. However, due to their black-box nature, they have brought several challenges in accepting the decisions derived from these models.

In addition, LLMs are also utilized in medical education to facilitate students and professionals in understanding complicated clinical ideas. The current review of the literature considers recent developments in the application of LLMs in healthcare, outlines their appropriateness, and accentuates the ethical and practical limitations involved.

11.2 PROPOSED ARCHITECTURE

The architecture shown in Figure 11.1 depicts a system intended to generate personalized recommendations for patients with diabetes. The system starts with the collection of data from two sources, namely the "Primal diabetes datasets" and general "Diabetes datasets." This combined data is stored and subsequently used for pretraining the BioGPT model-a specialized language model for biomedical text. Further, through chatbots, the system interacts with the patients and potentially uses Optical Character Recognition to recognize any text, with such interaction data being stored in a MongoDB database. A sample data is shown in Figure 11.2.

The next step involves creating specific prompts using data collected from patient interactions and OCR, which reside in the MongoDB database. These prompts serve as queries in the system or seek to check for diabetes indicators based on some specific information related to the patient. This prompt, being one input alongside the pretrained BioGPT model, is used in a fine-tuning process. The outcome-Fine-tuned model is, therefore, able to produce personalized recommendations for the patient based on specific diabetes types. This is delivered as an output, probably in a chart or report form.

DOI: 10.1201/9781003630241-14

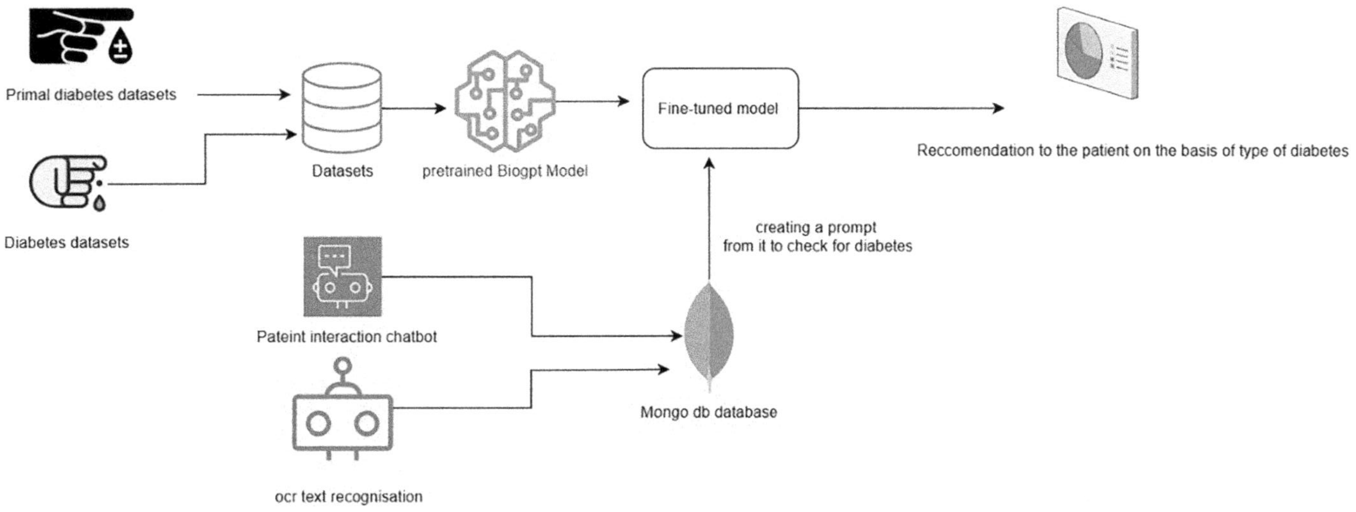

FIGURE 11.1 Architectural diagram of our multimodal data chatbot.

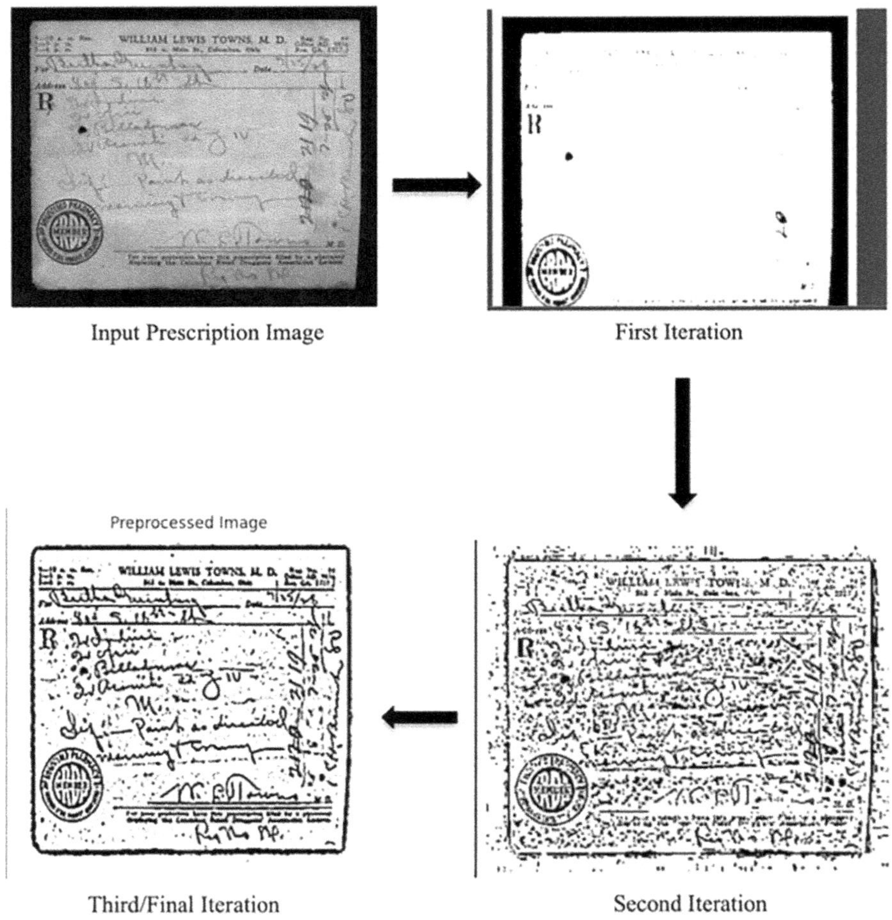

FIGURE 11.2 Sample taken for image processing.

11.3 PROPOSED METHODOLOGY

The proposed methodology follows a structured approach including data collection, processing, predictive modeling, and real-time alerting. Additionally, a web-based platform will be developed to provide healthcare professionals with explanations on how the decision is being made.

11.3.1 Data Preprocessing

The initial work was to put all raw data into editable and standard formats. Audio files were converted into mono channel at a sample rate of 16 kHz to satisfy the input requirements of the speech-to-text model.

As depicted in Figure 11.2, prescription images were improved by applying noise reduction filters, increasing contrast, and resizing them accordingly to improve readability for OCR models.

For tabular data, columns were standardized with correct naming conventions, missing values were imputed using appropriate techniques, and categorical responses such as "Yes" or "No" were converted to binary numerical values for easier model processing. The results obtained by reading the image can be seen in Figure 11.3.

Even after extensive preprocessing of the image, there was still a lot of noise, which was eliminated by applying a different OCR technique called Tesseract-OCR that produced the desired output. As can be seen in Figure 11.3, the application and result of Tesseract-OCR are shown in Figure 11.4.

```
Filtered Biochemistry Results (for Diabetes):
- HBA1C / GLYCOSYLATED
- HbA1C 5 4.0-6.0 %
- Estimated Average Glucose 96.80 90 - 120 mg/dL
- HbA1C % Result
```

FIGURE 11.3 Results by reading image.

flabs

hello@flabs.in
+91 7619837301

Plot No-20, Knowledge Park 2,IIT Roorkee Extension Centre, Tides Incubator, Greater Noida

Name	: Mr. Ayush	Patient ID	: 240329001
Age/Gender	: 23 Years/Male	Report ID	: RE378
Referred By	: Self	Collection Date	: 03/05/2024 10:41 AM
Phone No.	: 6397228160	Report Date	: 03/05/2024 10:43 AM

BIOCHEMISTRY
HBA1C / GLYCOSYLATED

TEST DESCRIPTION	RESULT	FLAG	REF. RANGE	UNIT
HbA1C	5		4.0 - 6.0	%
Estimated Average Glucose	96.80		90 - 120	mg/dL

Interpretation:

HbA1C %	Result
>8%	Action Suggested
7-8%	Good Control
<7%	Goal
6-7%	Near Normal Glycemia
<6%	Normal level

~~End of report~~

FIGURE 11.4 Scanned report.

11.3.2 Exploratory Data Analysis

Exploratory data analysis was performed before building any models to learn about the structure and quality of the datasets, as shown in Figure 11.5.

Summary statistics were cross-checked on tabular data, random samples of audio files were manually verified against background noise and clarity, and visual inspection of medical prescription images was performed to verify blur or corruption. This was a necessary step to identify inconsistencies such as mislabeled files or unreadable data that could affect model performance in the future.

11.3.3 Model Selection and Training

Depending on the data categories and activities to be performed, different tools and models were chosen. Random Forest for structured data, which among other things predicts appointment adherence, is a strong and interpretable model. For the task of speech-to-text conversion, the Whisper model was used, specifically its variant fine-tuned for Indian-accented Hinglish. Tr-OCR was used to extract text from handwritten prescriptions since it is intended for handwritten text recognition and conversion to digital format.

11.3.4 Fine-Tuning

All AI models were employed in their pretrained state to optimize time and resources. However, where required, domain-specialized variants were chosen to better suit the context, e.g., speech models fine-tuned for Indian dialects. Intermediate outputs such

```
import pandas as pd

# Assuming 'get_recommendations' is defined and returns a DataFrame
def get_recommendations(diabetes_type, age, blood_pressure, gender):
    # ... (Your existing get_recommendations logic) ...
    return recommendations_df  # Assuming a DataFrame is returned

# Get user inputs (from cell 60)
diabetes_type_input = input("Enter Diabetes Type: ").strip().lower()
age_input = int(input("Enter Age: "))
blood_pressure_input = int(input("Enter Blood Pressure Issue (1 for Yes, 0 for No): "))
gender_input = input("Enter Gender (Male/Female): ").strip().title()

# Get recommendations
recommendations_df = get_recommendations(diabetes_type_input, age_input, blood_pressure_input, gender_input)

# Save to text file
recommendations_df.to_csv('recommendations.txt', index=False)

# Download the file (Colab specific)
from google.colab import files
files.download('recommendations.txt')
```

```
Enter Diabetes Type: type 1
Enter Age: 65
Enter Blood Pressure Issue (1 for Yes, 0 for No): 1
Enter Gender (Male/Female): male
```

FIGURE 11.5 Exploratory data analysis.

```
# Step 8: Fine-tune the Model (Modified for T4 GPU)

from transformers import TrainingArguments, Trainer

training_args = TrainingArguments(
    output_dir="./finetuned_biogpt",           # Where model checkpoints will be saved
    num_train_epochs=3,                        # You can increase this later if needed
    per_device_train_batch_size=2,             #  Reduced for T4 GPU (was 4)
    gradient_accumulation_steps=2,             #  To simulate effective batch size of 4
    save_steps=500,
    save_total_limit=2,
    logging_steps=50,
    evaluation_strategy="no",
    remove_unused_columns=False,
    report_to="none",
    fp16=True,                                 # ✅ Enabled mixed precision for GPU
    logging_dir="./logs"                       # ✅ Optional: to store logs for tracking
)

trainer = Trainer(
    model=model,
    args=training_args,
    train_dataset=dataset,
    tokenizer=tokenizer
)

trainer.train()

# ✅ Save the fine-tuned model
model.save_pretrained("./finetuned_biogpt")
```

FIGURE 11.6 Code to take user information to register.

as audio transcriptions or extracted text from images were cached locally to avoid repetitive processing and to enhance the system's efficiency in future runs, as shown in Figure 11.6.

11.3.5 Integration with LLMs

The processed data from different sources were combined for in-depth analysis with the help of natural language models. This blend made contextual interpretation of patient information from varied inputs possible. Specific user information relating to his or her diabetic condition would be solicited from the user. This includes the user's type of diabetes (say for example, type 1), their age, if they have a blood pressure problem (1=yes; 0=no), plus the gender. After collecting that information, all of it is passed as arguments to a function called get_recommendations. This function (of which internals are not provided herein but assumed to exist within the project's context) processes these inputs and then creates certain recommendations, probably returned as a pandas DataFrame. Finally, the code saves these recommendations as recommendations.txt and includes a command to download the same using a method applicable to Google Colab.

The collected inputs, especially the user-specified diabetes_type, will hence feed into the system for specific recommendation generation. The type of diabetes entered together with the state of other health parameters such as age and blood pressure condition hence makes it possible for the system to recommend solutions or management strategies.

11.3.6 Connecting with MongoDB Atlas

For backend purposes, a MongoDB database called "Diabeticanfriend" has been implemented, containing collections such as biochemistry results.

The data represented in the screenshot show data extracted from biochemistry reports: details such as image names, test lines such as "RANDOM BLOOD SUGAR TEST," image paths, and processed image data may be stored in documents in this collection in a more structured way. This database acts as the central repository, and a sample is provided in Figure 11.7.

Data are written to this MongoDB collection, retrieved back from the database, and used by the recommendation system to further generate and display suitable recommendations based on the stored patient or test particulars.

System performance was compared using standard metrics. Accuracy and related statistical measures were employed for testing predictive models.

Transcriptions were checked by hand against original audio recordings to ensure quality; the results are mentioned in Figures 11.8 and 11.9.

FIGURE 11.7 Connecting to MongoDB *evaluation*.

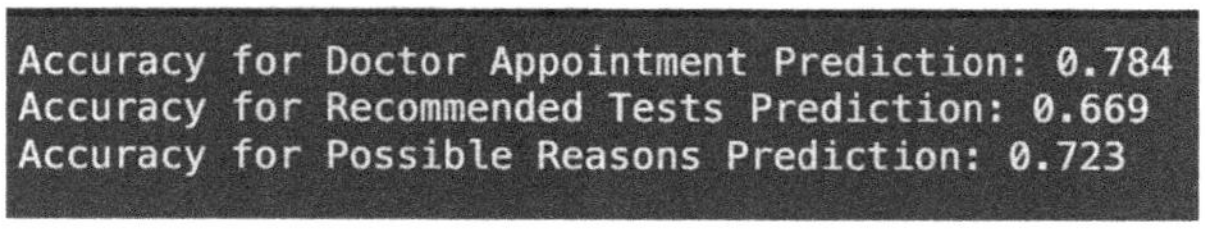

FIGURE 11.8 Accuracy for chatbot.

```
[10] # Evaluate model performance
     accuracy = accuracy_score(y_test, y_pred)
     print(f'Model Accuracy: {accuracy:.2f}')

     Model Accuracy: 0.52
```

FIGURE 11.9 Accuracy for documentation model.

11.3.7 Tackling Healthcare Staff Shortages

With growing administrative workloads and physician burnout, LLMs present a solution to avoid unnecessary work and optimize resource allocation in healthcare. Clinician exhaustion can be eased considerably through AI-driven decision-making and document automation.

11.3.7.1 Scaling Up Access to Healthcare Services

Telemedicine and AI-powered virtual assistants can enhance access to healthcare for marginalized populations, facilitate effective triage systems, symptom checkers, and remote consultations, thereby relieving congestion in healthcare centers.

11.3.7.2 Ensuring Ethical AI Deployment in Healthcare

As there can be biased AI decisions, there should be regulatory mechanisms in place that provide assurance of ethical deployment of LLMs. There must be clear AI models, understandable results, and human supervision to ensure trust in AI-enabled healthcare.

11.3.8 Deployment

Figures 11.10 and 11.11 show the frontend of our website, which is made with the help of Streamlit, where we can add reports, interact with the model, and also have conversations for decision support and recommendations about diabetes.

The assessment report using the LLM model is generated as shown in Figure 11.12 as per our model recommendation system.

The assessment report for female Kashish finds the weight healthy, reporting a BMI of 25.0. At present, she is "No diabetes (curious)" in the report, which names "Blood Pressure" and "Joint Pain" as the risk factors that need special attention. Kashish is also positively noted as being described as "Lightly Active" and following a "Low-Carb" diet, both of which are recognized as healthier lifestyle choices. The personal recommendations are that Kashish should seek guidance from a healthcare professional for actual dietary advice; in other words, the current low-carb diet should be discussed with an expert.

The recommended exercise consists of a structured plan for 5 days a week for 30 minutes, which includes mild walking, swimming, light resistance training, and stretching. The report also recommends regular blood sugar checks, a healthy diet, keeping physically active, as well as attending regular check-ups.

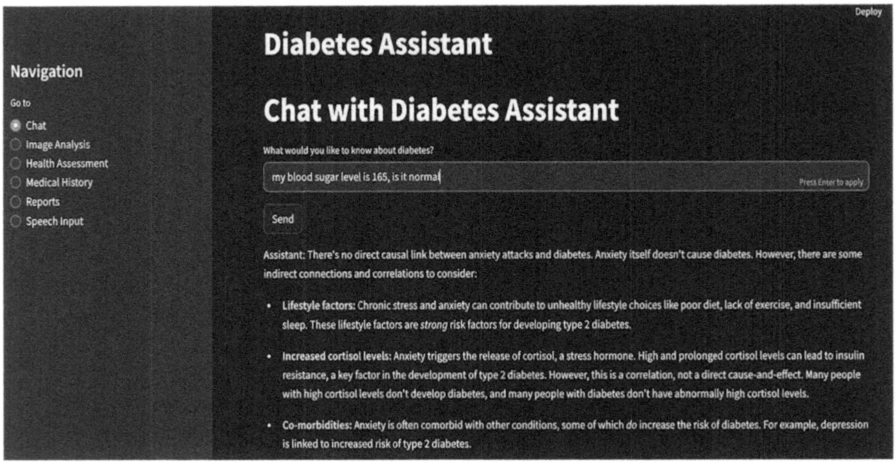

FIGURE 11.10 AI helping in healthcare.

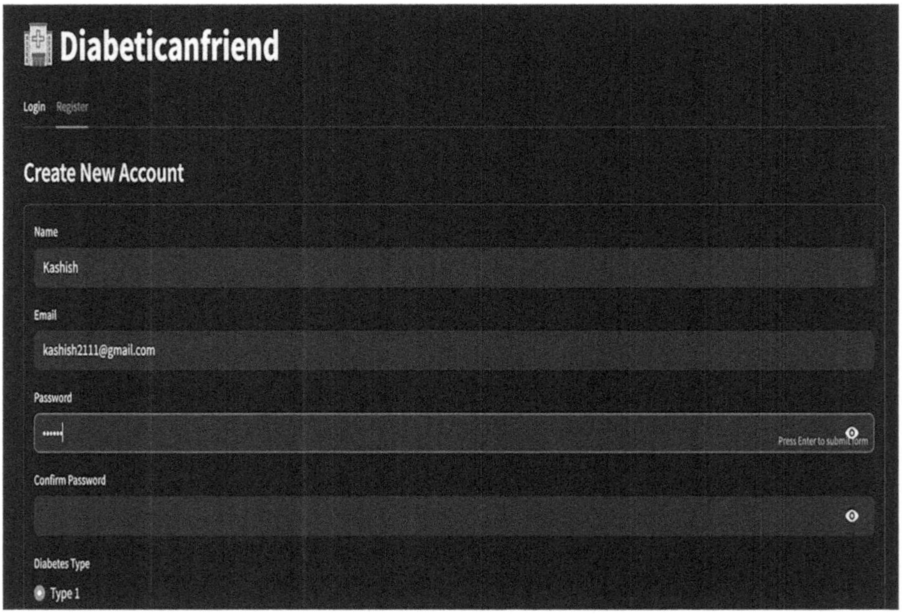

FIGURE 11.11 Frontend of the application.

Diabetes Health Assessment Report

Generated on: 2025-04-25 22:31:07

Patient Information

Name: Kashish
Email: kashish2111@gmail.com
Age: 19
Gender: Female
Diabetes Type: No diabetes (curious)
Medical History: anxiety attacks

Health Assessment Results

BMI: 25.0 (Normal)
Risk Factors:
• Blood Pressure
• Joint Pain
Lifestyle Factors:
• Activity Level: Lightly Active
• Diet Type: Low Carb

Personalized Recommendations

Diet Plan

• Consult with your healthcare provider for personalized diet recommendations

Exercise Plan

Frequency: 5 days per week
Duration: 30 minutes per session
Recommended Activities:
• Walking
• Swimming
• Light resistance training
• Gentle stretching

Lifestyle Recommendations

• Monitor blood sugar regularly
• Maintain a healthy diet
• Stay physically active
• Get regular check-ups

FIGURE 11.12 Assessment report.

11.4 SUMMARY

As these models are capable of reading and writing like humans, their contextually appropriate responses and decision accuracy have led to the models being utilized for highly critical healthcare applications with improved precision in documentation processes. However, due to their black-box nature, they have brought several challenges in accepting the decisions derived from these models.

In addition, LLMs are also utilized in medical education to facilitate students and professionals in understanding complicated clinical ideas. The current review of the literature considers recent developments in the application of LLMs in healthcare, outlines their appropriateness, and accentuates the ethical and practical limitations involved.

Hence, this chapter focuses on the application of XAI in the LLM model so that we can get insights on how the model works in real-time healthcare analytics.

BIBLIOGRAPHY

Aldwean, A., and Tenney, D. (2024). Large language models in healthcare: Bridging the gap between performance evaluation and socio-ethical implications. In 2024 ASEE North East Section.

AlSaad, R., Abd-Alrazaq, A., Boughorbel, S., Ahmed, A., Renault, M. A., Damseh, R., and Sheikh, J. (2024). Multimodal large language models in health care: Applications, challenges, and future outlook. *Journal of Medical Internet Research*, 26, e59505.

Andrew, A. (2024). Potential applications and implications of large language models in primary care. *Family Medicine and Community Health*, 12(Suppl 1), e002602.

Cascella, Marco, Semeraro, Federico, Montomoli, Jonathan, Bellini, Valentina, Piazza, Ornella, and Bignami, Elena. (2024). The breakthrough of large language models release for medical applications. *Journal of Medical Systems*, 48(1), 22.

Deng, Heng, and Zhou, Qian. (2024). *The Current Status and Prospects of Large Language Models in Medical Fields*. Springer.

Ferrario, A., and Biller-Andorno, N. (2024). Large language models in medical ethics: useful but not expert. *Journal of Medical Ethics*, 50(9), 653–654.

Gładysz, M., Fiumedinisi, F., Burn, F., Rommers, N., Giovanoli, P., and Plock, J. A. (2026). AI-Assisted Medical Documentation in a Multilingual Swiss Health Care System: Proof-of-Concept Study. *JMIR AI*, 5, e77351.

Haltaufderheide, Joschka, and Ranisch, Robert. (2024). The ethics of ChatGPT in medicine and healthcare: A systematic review on large language models (LLMs). arXiv.

He, Kai, Mao, Rui, Lin, Qika, Ruan, Yucheng, Lan, Xiang, Feng, Mengling, and Cambria, Erik. (2023). A survey of large language models for healthcare: From data, technology, and applications to accountability and ethics. arXiv.

Jung, H., Kim, Y., Choi, H., Seo, H., Kim, M., Han, J., Kee, G., Park, S., Ko, S., Kim, B., and Kim, S. (2024). Enhancing clinical efficiency through LLM: Discharge note generation for cardiac patients. arXiv.

Karttunen, Pinja. (2023). Large language models in healthcare decision support. Bachelor's Thesis.

Mirzaei, T., Amini, L., and Esmaeilzadeh, P. (2024). Clinician voices on ethics of LLM integration in healthcare: a thematic analysis of ethical concerns and implications. *BMC Medical Informatics and Decision Making*, 24(1), 250.

Omar, M., Nadkarni, G. N., Klang, E., and Glicksberg, B. S. (2024). Large language models in medicine: a review of current clinical trials across healthcare applications. *PLOS Digital Health*, 3(11), e0000662.

Oniani, David, Wu, Xizhi, Visweswaran, Shyam, Kapoor, Sumit, Kooragayalu, Shravan, Polanska, Katelyn, and Wang, Yanshan. (2024). Enhancing large language models for clinical decision support by incorporating clinical practice guidelines. arXiv.

Shi, Tongyue, Ma, Jun, Yu, Zihan, Xu, Haowei, Xiong, Minqi, Xiao, Meirong, Li, Yilin, Zhao, Huiying, and Kong, Guilan. (2024). Stochastic parrots or ICU experts? Large language models in critical care medicine: A scoping review. arXiv.

Vaid, A., Lampert, J., Lee, J., Sawant, A., Apakama, D., Sakhuja, A., Soroush, A., Lee, D., Landi, I., Bussola, N., and Nabeel, I. (2024). Generative large language models are autonomous practitioners of evidence-based medicine. arXiv.

Wang, L., Wan, Z., Ni, C., Song, Q., Li, Y., Clayton, E., Malin, B., and Yin, Z. (2024). Applications and concerns of ChatGPT and other conversational large language models in health care: Systematic review. *Journal of Medical Internet Research*, 26, e22769.

Wang, Dandan, and Zhang, Shiqing. (2024). *Large Language Models in Medical and Healthcare Fields: Applications, Advances, and Challenges*. Springer.

Wen, Bo, Norel, Raquel, Liu, Julia, Stappenbeck, Thaddeus, Zulkernine, Farhana, and Chen, Huamin. (2024). Leveraging large language models for patient engagement: The power of conversational AI in digital health. arXiv.

Yang, Rui, Marrese-Taylor, Edison, Ke, Yuhe, Cheng, Lechao, Chen, Qingyu, and Li, Irene. (2023). Integrating UMLS knowledge into large language models for medical question answering. arXiv.

Yang, Rui, Tan, Ting Fang, Lu, Wei, Thirunavukarasu, Arun James, Ting, Daniel Shu Wei, and Liu, Nan. (2023). *Large Language Models in Health Care: Development, Applications, and Ethical Considerations*. Wiley.

Yu, H., Fan, L., Li, L., Zhou, J., Ma, Z., Xian, L., Hua, W., He, S., Jin, M., Zhang, Y., and Gandhi, A. (2024). *Large Language Models in Biomedical and Health Informatics: A Review with Bibliometric Analysis*. Springer.

Zhang, K., Meng, X., Yan, X., Ji, J., Liu, J., Xu, H., Zhang, H., Liu, D., Wang, J., Wang, X., and Gao, J. (2025). Revolutionizing healthcare: The transformative impact of large language models in medicine. *Journal of Medical Internet Research*, 27, e59069.

Zhang, Xiaocheng, Yao, Zonghai, and Yu, Hong. (2023). EHR interaction between patients and AI: NoteAid EHR interaction. arXiv.

Zou, Shun, and He, Jun. (2023, October). Large language models in healthcare: A review. In *Proceedings of the 7th International Symposium on Computer Science and Intelligent Control (ISCSIC)* (pp. 141–145). IEEE.

Index

Pages in *italics* refer to figures, pages in **bold** refer to tables, and pages followed by n refer to notes.

K

L

M

For Product Safety Concerns and Information please contact our EU
representative GPSR@taylorandfrancis.com
Taylor & Francis Verlag GmbH, Kaufingerstraße 24, 80331 München, Germany

www.ingramcontent.com/pod-product-compliance
Ingram Content Group UK Ltd.
Pitfield, Milton Keynes, MK11 3LW, UK
UKHW021835091026
476426UK00003B/46
* 9 7 8 1 0 4 1 0 4 6 8 7 5 *